Beiträge zur organischen Synthese

Band 98

Beiträge zur organischen Synthese

Band 98

Herausgegeben von
Prof. Dr. Stefan Bräse

Karlsruher Institut für Technologie (KIT)
Institut für Organische Chemie
Fritz-Haber-Weg 6, D-76131 Karlsruhe

Institut für Biologische und Chemische Systeme – Funktionelle Molekulare Systeme
Hermann-von-Helmholtz-Platz 1
D-76344 Eggenstein-Leopoldshafen

Gloria Hong

Design and Synthesis of Emissive and Switchable Photo-Active Molecules

Logos Verlag Berlin

λογος

Bibliographic information published by the Deutsche Nationalbibliothek

The Deutsche Nationalbibliothek lists this publication in the Deutsche Nationalbibliografie; detailed bibliographic data are available in the Internet at http://dnb.d-nb.de
.

ISBN 978-3-8325-5624-2
ISSN 1862-5681

Logos Verlag Berlin GmbH
Georg-Knorr-Str. 4, Geb. 10, 12681 Berlin

Tel.:   +49 (0)30 / 42 85 10 90
Fax: +49 (0)30 / 42 85 10 92

https://www.logos-verlag.de

Therefore we do not lose heart.
Though outwardly we are wasting away,
yet inwardly we are being renewed day by day.

So we fix our eyes not on what is seen,
but on what is unseen,
since what is seen is temporary,
but what is unseen is eternal.

2 Corinthians 4:16, 18 (NIV)

Table of Contents

German Title of the Thesis

Design und Synthese von emittierenden und schaltbaren, photoaktiven Molekülen

Kurzzusammenfassung

Photoaktive organische Moleküle wie Luminophore und Photoschalter weisen nützliche Eigenschaften wie das Ausstrahlen von Licht einer definierten Wellenlänge oder einen Konformationswechsel auf, die in unterschiedlichsten Anwendungen eingesetzt werden können. Luminophore werden als Emitter in organischen Leuchtdioden (OLEDs) oder als Farbstoff in zellulären Bildgebungsverfahren genutzt. Die Suche nach effizienten Emittern, die umweltschädlich produzierte Schwermetalle vermeiden, hat zu der Entwicklung von rein organischen Materialien geführt, die thermisch aktivierte, zeitverzögerte Fluoreszenz (TADF) aufweisen. Die Entwicklung von effizienten, roten TADF Emittern bleibt herausfordernd. Daher wurden in dieser Arbeit drei elektronenakzeptierende Moleküleinheiten für rote TADF Emitter untersucht. Hierbei wurden Syntheserouten für die Darstellung von Dibenzo[a,c]phenazin (BP)-basierten Donor (D) und Akzeptor (A) Konjugaten der folgenden Typen etabliert: D-A, 2D-A-D, 2D-A-R. Die grün bis rot emittierenden BP-basierten Materialien wiesen Emissionswellenlängenmaxima bis zu 716 nm auf und die TADF Eigenschaften der 2D-BP-F Serie konnten experimentell bestätigt werden. Die grün- bis rot-emittierenden OLED Geräte der 2D-BP-F Serie demonstrierten exzellente externe Quanteneffizienzen (EQE) von bis zu 21.8%. Darüber hinaus wurden zwei neue Akzeptoren basierend auf Pyrrolo[3,4-f]isoindole-1,3,5,6($2H$, $6H$)-tetraon (PIT) und 1H-Pyrrolo[3,4-b]quinoxaline-1,3($2H$)-dion (PQD) entwickelt. Die Synthese eines PIT-basierten Emitters wurde nach mehrmaliger Iteration des Zielmoleküldesigns realisiert. Im Falle des PQD Akzeptors wurde die Derivatisierung an der Imid-Position und unterschiedliche Verknüpfungsmöglichkeiten der Donor- und Akzeptoreinheit untersucht. Die PQD-basierten Emitter wiesen orange bis rote Emission mit Wellenlängenmaxima bis zu 677 nm auf. Im Gegensatz zu Luminophoren, reagieren Photoschalter auf die Absorption von Licht mit einer Änderung der Konformation zu einem entsprechenden Isomer. Azobenzole gehören zu den prominentesten Vertretern dieser Materialien und können in Verbindung mit Liganden wie (−)-Menthol, die optische Kontrolle von TRPM8 Ionenkanälen ermöglichen. Diese Arbeit hat sich mit der Synthese von modifizierten Azobenzolen mit Photoschaltbarkeit bei längeren Wellenlängen und verschiedenen Verknüpfungen zur (−)-Menthyl-Einheit befasst. Zuletzt wurde ein Azobenzol-basiertes Push-Pull-System synthetisiert, um eine Donor- und Akzeptoreinheit nach dem Photoschalten in räumliche Nähe zu bringen.

Abstract

Photoactive organic molecules such as luminophores and photoswitches show useful properties such as the emission of light of a defined wavelength and conformational changes upon irradiation with light, which are advantageous for various applications. Luminophores, for example, are readily employed as emitters in organic light-emitting diodes (OLEDs) or as dyes in cellular imaging. The investigation of efficient emitters that do not require environmentally harmful sourced materials, such as heavy metals, has led to the development of purely organic thermally activated delayed fluorescence (TADF) materials.[1] The challenge of obtaining highly efficient red TADF emitters, amidst the energy gap law, remains. Thus, three aromatic, heterocyclic electron-accepting systems, henceforth called "acceptors", for red-emitting TADF molecules were investigated in this work. The establishment of synthesis routes towards dibenzo[a,c]phenazine (BP)-based donor (D)-acceptor (A) conjugates of the following types were successful: D-A, 2D-A-D and 2D-A-R. BP-based green to red-emitting materials with emission wavelength maxima of up to 716 nm were obtained, and TADF properties were experimentally confirmed for the 2D-BP-F series. The OLED devices featuring these emitters showed maximum external quantum efficiencies (EQEs) of up to 21.8%. Furthermore, two novel acceptors, namely pyrrolo[3,4-f]isoindole-1,3,5,6($2H$, $6H$)-tetraone (PIT) and $1H$-pyrrolo[3,4-b]quinoxaline-1,3($2H$)-dione (PQD), were designed and the syntheses of donor and acceptor conjugates were successfully carried out. While the PIT-based system underwent several iterations of target structure adaptation according to synthetic challenges faced along the way, the PQD acceptor was investigated regarding derivatization at the imide position and the connecting position of the donor unit at the acceptor core. The PQD-based emitters showed orange to red emissions with wavelengths of up to 677 nm. On the other hand, photo-responsive materials such as azobenzenes form one of the most prominent classes of photoswitches. Attached to ligands such as menthol, they can be used to gain optical control over TRPM8 ion channels. In this work, azobenzenes with modified substitution patterns were synthesized to achieve photoswitching at longer wavelengths, and different connectivities to the (−)-menthyl unit were investigated. Lastly, a push-pull system based on an azobenzene structure was designed and synthesized to bring a donor and acceptor moiety into spatial proximity upon photoswitching.

1. Introduction

Organic molecules find applications as pharmaceuticals, materials, catalysts, and many more. While some research on organic chemistry is devoted to investigating underlying reaction mechanisms and developing new reaction types, other areas of organic chemistry are more application-oriented and serve to synthesize molecules bearing defined qualities or features that are required for a specified purpose. For example, organic molecules that show thermally activated delayed fluorescence (TADF) are readily employed as luminescent emitters in optoelectronics, e.g., in organic light-emitting diodes (OLEDs)[2], but also attract increasing interest as metal-free luminophores for biomedical applications as highlighted in a recent review by Zhang and coworkers.[3] Other organic molecules depict the ability to change their conformation upon irradiation with light – a property known as "photoswitching". Their use is as versatile as organic fluorophores. Their ability to offer an optical switch is appreciated in photopharmacology, where the control of the biological activity of ligands is highly desirable concerning the spatial and temporal selectivity in drug action.[4-5]

Organic TADF fluorophores for OLED materials offer decisive benefits. Firstly, they do not require the use of rare noble metals harvested through environmentally damaging processes. Furthermore, thermally activated delayed fluorescence (*vide infra*) allows for up to 100% internal quantum efficiencies, leading to high external quantum efficiencies in OLED devices. A recent review by the Bräse group gave an overview of emitters that were developed for OLEDs.[2] In an OLED stack, emitting materials such as TADF molecules are employed in the emissive layer, as shown in Figure 1.

In general, the OLED stack is built upon a transparent substrate, upon which the transparent anode is located. A typical anode material is indium tin oxide (ITO, $(In_2O_3)_{0.9}(SnO_2)_{0.1}$). Stacked on top of this electrode, the hole injection layer (HIL) is placed, followed by a hole transport layer (HTL), which delivers the positive charges, or holes, to the EML. Here, the recombination of holes and electrons leads to the formation of excitons. The electrons are supplied *via* the electron transport layer (ETL) and the electron injection layer (EIL) adjacent to the cathode.[6]

14

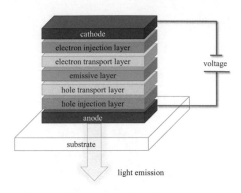

Figure 1. Schematic visualization of an OLED stack architecture.

Beyond their OLED application, TADF fluorophores offer attractive options for bioimaging purposes. Luminophores emitting prompt fluorescence in the red spectral region above 620 nm assist in avoiding the autofluorescence of cells that is typically detected at shorter wavelengths. Thus, the signal-to-noise ratio is measurably improved.[7] Furthermore, the delayed fluorescence component that is characteristic for TADF can be used for time-resolved luminescence imaging (TRLI), for which fluorophores with long excited-state lifetimes of $\tau > 10$ ns can remove the short-lived background fluorescence.[8] However, some challenges remain, such as oxygen sensitivity which is connected to quenching the triplet state through the formation of singlet oxygen $^1O_2$, and water solubility, amongst others.[3]

A challenge to one application is the asset of another application field. The production of $^1O_2$ through the quenching of the triplet state in TADF molecules can be used in photodynamic therapy (PDT) for cancer treatment. PDT is a highly promising method as it is non-invasive and shows high temporal-spatial resolution and minimal drug resistance and side effects.[9-13] In PDT, the formation of cytotoxic and reactive singlet oxygen $^1O_2$ by energy transfer to a ground state triplet oxygen $^3O_2$ (type II PDT) and other reactive oxygen species (ROS) (type I PDT) is desired to eliminate tumor cells by inducing apoptosis and/or necrosis.[12, 14-15] For the formation of $^1O_2$ and ROS, the malignant tissues or cells are exposed to a photosensitizer (PS). Then, they are ideally irradiated with visible light or light in the red/near-infrared region. Thereby, the PS is activated and forms an excited triplet state. Relaxation of the PS is subsequently induced through energy transfer to either ground-state oxygen, generating the desired $^1O_2$ or undergoing a one-electron oxidation-reduction reaction with a neighboring molecule that leads to free radical intermediates reacting with oxygen to produce peroxy radicals and other ROS.[14]

Molecules that show thermally activated delayed fluorescence are diverse and possess intriguing features that cater to various applications. These will be discussed in more depth in the following subchapter.

1.1. Thermally Activated Delayed Fluorescence

The feature that makes TADF molecules a promising source of novel materials is based on the photophysical processes upon energy absorption. In this subchapter, the basic underlying photophysics and variables of importance are introduced. The Jablonski diagram shown in Figure 2 offers an appropriate starting point as it schematically visualizes the photophysical processes upon energy input into a system.

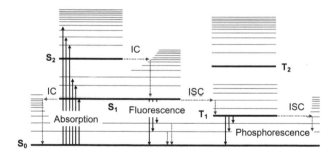

Figure 2. Schematic Jablonski diagram as seen in *Molecular Fluorescence: Principles and Applications* by B. Valeur, chapter 3.1, page 35; IC: internal conversion, ISC: intersystem crossing, vertical dotted arrows indicating vibrational relaxation.[16]

As a photon is absorbed, an electron is promoted from a molecular orbital of the ground state to an unoccupied molecular orbital to form an excited state. This electronic transition can occur between different types of molecular orbitals and the energy of these electronic transitions can be generalized to follow this order: $n \rightarrow \pi^* < \pi \rightarrow \pi^* < n \rightarrow \sigma^* < \sigma \rightarrow \pi^* < \sigma \rightarrow \sigma^*$.[16] Once in the excited state, the system relaxes to the ground state through de-excitation processes. Deactivation of the excited state occur through radiative pathways, vibrational relaxation, and other non-radiative pathways, as shown in Figure 2. Vibrational relaxation (VR) describes the dissipation of energy by interaction with neighboring molecules. In addition to intramolecular vibrational redistribution, which characterizes the energy distribution to other vibrational modes, they are the most rapid processes that take place in the excited state. Due to transitions between isoenergetic vibrational levels of different electronic states, deactivation occurs *via* radiationless transitions such as internal conversion (IC) and intersystem crossing (ISC). While IC describes the radiationless transition between states of equal spin multiplicity, e.g., $S_n \rightarrow S_1$, ISC occurs between states of different spin multiplicity, e.g., $S_1 \rightarrow T_1$. The latter is a spin-forbidden transition, but it is made possible due to spin-orbit coupling (SOC). Lastly, deactivation *via* radiative pathways leads to the relaxation to the ground state with the emission of a photon (luminescence). Following Kasha's rule stating that polyatomic molecular luminescence is observed only from the lowest excited state of a given multiplicity, only the lowest-lying excited singlet state S_1 and lowest-lying excited triplet state T_1 are usually considered in this context.[17-18]

Three types of radiative de-activation will be differentiated in the following. Firstly, the transition from $S_1 \rightarrow S_0$ is called fluorescence, while secondly, the $T_1 \rightarrow S_0$ transition is known as phosphorescence. As phosphorescence implies the transition between two states of different multiplicity, it is a spin-forbidden process enabled *via* SOC.[16, 18] The third luminescence type is called thermally activated delayed fluorescence and is observed in molecules, where the energy difference ΔE_{ST} between S_1 and T_1 is small. In molecules displaying TADF properties, a non-radiative process called reverse intersystem crossing (RISC) is enabled through thermal activation at room temperature. It leads to the upconversion of triplet excitons to the S_1 state, from which they then relax to show delayed fluorescence besides the prompt fluorescence. The RISC allows for high internal quantum efficiencies (IQEs) up to 100% as both the singlet and triplet excitons can be harvested. The rate constant for RISC is correlated to the ΔE_{ST} as follows, whereas k_B is the Boltzmann constant and T is the absolute temperature:[19]

$$k_{RISC} \propto exp\left(-\frac{\Delta E_{ST}}{k_B T}\right) \tag{1}$$

As depicted in equation (1), the rate constant of RISC increases upon the reduction of the energy difference ΔE_{ST}. ΔE_{ST} is given by equation (2) in dependence of the exchange integral J.[20]

$$\Delta E_{ST} = E_{S1} - E_{T1} = 2J \tag{2}$$

The exchange integral is defined as shown in equation (3), whereas $\phi(r)$ and $\psi(r)$ are the wave functions of the highest occupied molecular orbital (HOMO) and lowest unoccupied molecular orbital (LUMO), and e the electron charge.[20]

$$J = \iint \phi(r_1)\psi(r_2) \left(\frac{e^2}{r_1 - r_2}\right) \phi(r_2)\psi(r_1) dr_1 dr_2 \tag{3}$$

The above equation is applicable when the excited states are pure HOMO-LUMO transitions. The correlation between equations (1), (2), and (3) shows that a smaller overlap between the HOMO and LUMO leads to a reduction of the exchange integral J and, therefore, a diminished ΔE_{ST}. In terms of molecular design, these frontier molecular orbitals (FMOs) can be spatially separated by installing electron-donating (D) and electron-accepting (A) moieties that prefer D-A electron transfer from the HOMO to the LUMO, respectively, in excited states with strong intramolecular charge transfer (ICT) character. Moreover, the energy gap between the excited singlet and triplet state is minimized by introducing a twist between D and A or using a spacer that increases the distance between D and A, as reported by Gierschner and Milián-Medina.[21] However, in their work, they also note that a too-small ΔE_{ST}, and thereby a negligible FMO overlap is disadvantageous as the oscillator strength f (*vide infra*) will converge to zero, which leads to a depressed rate constant of fluorescence, k_F (*vide infra*), which then results in a low photoluminescence quantum yield (PLQY, *vide infra*). In case the excited states are not pure HOMO-LUMO transitions conducted by ICT, many-body electronic wavefunctions for the singlet and triplet states take the place of $\phi(r)$ and $\psi(r)$ in equation (3), and the spatial separation of HOMO and LUMO alone is insufficient to minimize ΔE_{ST}.[20, 22] The HOMO and LUMO can be experimentally approximated using cyclic voltammetry (CV). In addition to a strategy based on highly twisted structures, through-space conjugation and charge transfer were investigated for the design of TADF molecules.[23-26]

On the assumption of the classical theory that a molecule can be understood as an oscillating dipole, the oscillator strength f is a parameter that describes the molecular absorption of light and is correlated to the molar extinction coefficient $\varepsilon(\lambda)$ [$M^{-1}cm^{-1}$], the index of refraction n and the wavenumber $\bar{\nu}$ [cm^{-1}] as follows:[16]

$$f = \frac{4.32 \cdot 10^{-9}}{n} \int \varepsilon(\bar{v}) \, d\bar{v} \qquad (4)$$

To put values for the dimensionless oscillator strength f into perspective, the oscillator strengths for known organic compounds such as benzene, acridine, and Eosin Y are approximately 0.2, 12, and $90 \cdot 10^3$ M^{-1} cm^{-1}, respectively.[16]

The molar extinction coefficient $\varepsilon(\lambda)$, which defines the oscillator strength, is a parameter that is experimentally obtained through the measurement of absorbance. Absorbance A measures light absorption efficiency at a specific wavelength λ [nm]. The Lambert-Beer law as shown in equation (5), describes the absorbance A of a sample, whereas $\varepsilon(\lambda)$ represents the molar extinction coefficient, c [M] the concentration, and l [cm] the absorption pathway length.

$$A (\lambda) = \varepsilon(\lambda) \cdot c \cdot l \qquad (5)$$

Data obtained from absorption measurements is often visualized as a plot of the molar extinction coefficient $\varepsilon(\lambda)$ against the wavelength λ, as $\varepsilon(\lambda)$ indicates the ability to absorb light in a given solvent.[16]

A higher oscillator strength increases the rate constant for fluorescence, k_F, according to the correlation shown in equation (6), whereas e and m are the charge and mass of an electron, respectively, and c is the speed of light, ε_0 is the emission coefficient and ω and f are the S_1-S_0 transition frequency and oscillator strength, respectively.[27]

$$k_F = \frac{e^2 \omega_{10}^2}{2\pi\varepsilon_0 mc^3} \cdot f_{10} \qquad (6)$$

Dependent on the rate constant of fluorescence is the photoluminescence quantum yield (PLQY, ϕ). The PLQY is defined in dependency on the respective rate constants k of fluorescence (F), internal conversion (IC) and intersystem crossing (ISC), as depicted in the following equation (7):[28]

$$PLQY = \phi_F = \frac{k_F}{k_F + k_{IC} + k_{ISC}} \qquad (7)$$

The PLQY for TADF molecules maximizes as the rate constant of internal conversion k_{IC} minimizes. Internal conversion describes a nonradiative transition between electronic states of the same spin multiplicity and thereby displays an unwanted process in light-emitting devices. In blue- or green-emitting molecules, the energy gap between S_1 and S_0 is usually large enough to suppress above mentioned nonradiative transitions of the S_1 excitons.[29] However, in red emitters that emit longer wavelengths of lower energy, the energy gap between S_1 and S_0 is sufficiently small to enable IC, resulting in a higher probability of a decreased PLQY. Therefore, the molecule design of (deep-)red emitters remains challenging and is a fine balancing of emission wavelength and photoluminescence quantum yield.

Lastly, the external quantum efficiency (EQE, η_{ext}) is one of the most indicative variables regarding the successful application of TADF emitters in OLEDs. It is defined as described in equation (8), whereas η_{int} and η_{out} are the internal quantum efficiency and out-coupling efficiency, respectively, γ is the ratio of the charge injection to the electron and hole transportation, and the indices d and p are the nominators for the delayed and prompt fluorescence.[30]

$$\eta_{ext} = \eta_{int} \cdot \eta_{out} \tag{8}$$

$$= \gamma \cdot \left(0.25 \cdot \phi_p + \left(0.75 + 0.25 (1 - \phi_p) \right) \frac{\phi_d}{1 - \phi_p} \right) \cdot \eta_{out}$$

In organic light-emitting diodes, 25% of the generated excitons are in the excited singlet state, while 75% are in the excited triplet state.[31] As TADF uses upconversion *via* RISC to harvest the triplet excitons, internal quantum efficiencies up to 100% are theoretically possible. The EQE can be increased while the outcoupling efficiency stays constant.

Next to the experimental characterization of TADF compounds, the ground state geometry and excited states can be calculated by computational methods based on density functional theory (DFT) and time-dependent DFT (TD-DFT).[32-33] Ground state geometry optimization is especially helpful in gaining information on where the highest occupied molecular orbital (HOMO) and the lowest unoccupied molecular orbital (LUMO) are located. For calculating the excited state, which involves a significant amount of charge-transfer (CT), the TAMM-DANCOFF approximation (TDA) is used to aid the computation of the triplet excitation energies.[34]

1.2. Molecule Design of TADF Fluorophores

The molecule design of TADF fluorophores is as diverse as its applications. The design rationale can be tailored to match the needs of different applications. Motivations for TADF design strategies are to develop materials that emit at a precise wavelength, show a certain efficiency (e.g., in OLEDs), feature biocompatibility (e.g., for cellular imaging), or act as linker materials in structures of higher degrees of order, etc. The design behind TADF materials for specific purposes is explored in the following subchapters.

1.2.1. Orange-Red TADF Emitters for OLEDs

The first orange-red TADF emitter for OLEDs, namely **4CzTPN-Ph (1)**, is shown in Figure 3 and was reported by Adachi and coworkers in 2012.[35] When built into an OLED device, this emitter showed an EQE of 11.2 ± 1% and an emission wavelength of 577 nm in toluene. Ever since, the development of (deep-)red to near-infrared TADF fluorophores has flourished.[36] The design of (deep-)red TADF emitters poses an inherent challenge. As shown in equation (7), the PLQY is not only dependent on the rate constant of radiative decay processes such as fluorescence but also on non-radiative de-excitation through internal conversion (IC) and intersystem crossing (ISC). Englman and Jortner reported that in large, aromatic molecules, where the electronic relaxation lies within the rule of a weak coupling limit, the rate constant of the non-radiative decay is inversely proportional to the exponential of the optical energy gap ΔE_{opt}.[37-38] On the other hand, the rate constant of the radiative decay is only proportional to the cube of ΔE_{opt}.[39] Therefore, the influence of non-radiative processes on the photoluminescence quantum yield increases exponentially as the energy of the emissive excited state decreases because of the vibronic coupling between the excited state and ground state, which is facilitated.[38] One method to reduce losses due to vibrational quenching and other non-radiative decay pathways in TADF molecules emitting at longer wavelengths is the increased rigidity of the molecular structure of the donor and acceptor units.[40-41]

The dibenzo[*a,c*]phenazines (BPs) were a suitably rigid *N*-heterocycle used as an electron-accepting moiety in orange-red TADF emitters. As displayed in Figure 3, the emitter series **1DMAC-BP (2a)**, **2DMAC-BP (2b)**, and **3DMAC-BP (2c)**, as well as **1PXZ-BP (3a)**, **2PXZ-BP (3b)**, and **3PXZ-BP (3c)** that followed the (poly-)donor-acceptor design were reported by Zhao and coworkers.[42-43] In dependence on the number of donor units attached to the BP acceptor core, the 9,9-dimethyl-9,10-dihydroacridine (DMAC)-based emitter series displayed electroluminescence up to a wavelength maximum of 606 nm (**3DMAC-BP, 2c**) and PLQYs of up to 89% (**3DMAC-BP, 2c**). The ΔE_{ST} decreased from 0.22 eV (**1DMAC-BP, 2a**) to 0.05 eV (**3DMAC-BP, 2c**) with an increasing number of donors and was found to be low for all three emitters. The maximum EQE of this series was 22% for a device featuring **3DMAC-BP (2c)**.[43]

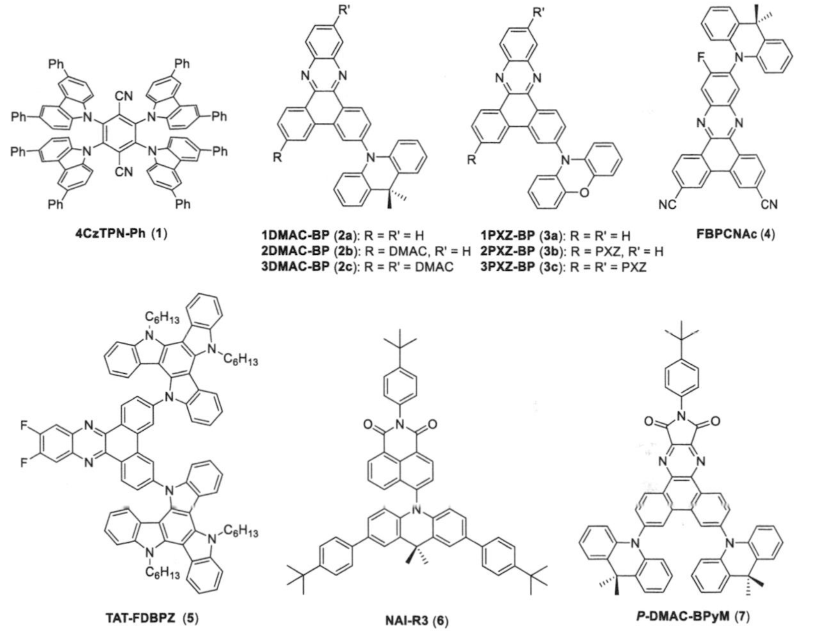

Figure 3. Dibenzo[a,c]phenazine- and aromatic imide-based TADF emitters for OLEDs.[35, 41-46]

The analogous series using the stronger donor unit 10H-phenoxazine (PXZ) showed a red-shifted maximum electroluminescence wavelength of 634 nm for an OLED device containing **3PXZ-BP (3c)**. The ΔE_{ST} for the PXZ-based series was in a similar range to the DMAC-based series and depicted values between 0.25 eV (**1PXZ-BP, 3a**) to 0.03 eV (**3PXZ-BP, 3c**). However, in contrast to the DMAC-base series, an increasing number of donor units led to a depression of the PLQY for the PXZ-based series. While for **1PXZ-BP (3a)**, a PLQY of 73% was observed, the PLQY for **3PXZ-BP (3c)** amounted to 22%. The photoluminescence quantum yield is a decisive factor for the EQE; this trend was also observed for the external quantum efficiency. **1PXZ-BP** (**3a**) displayed the highest maximum EQE of 26.3%, while **3PXZ-BP** (**3c**) showed an EQE_{max} of 7.1%.[42] The investigation of both BP-based emitter series gave insight into the scope and limitations of the (poly-)donor-acceptor design.

Lee and coworkers had investigated the importance of the rigid aromatic system for the BP-based TADF emitter **FBPCNAc** (**4**), which showed a small ΔE_{ST} of 0.047 eV, an electroluminescence wavelength of 597 nm, a PLQY of 79%, and an EQE_{max} of 23.8%.[41] Furthermore, their approach featured an auxiliary fluorine acceptor on the BP acceptor core. Wang and coworkers investigated the effect of fluorine substituents on the BP-acceptor and reported that the usage of fluorine substituents led to a deeper LUMO, which successfully shifted the emission to longer wavelengths. Furthermore, the ΔE_{ST} was slightly minimized, but these effects came at the cost of a reduced PLQY and EQE_{max}. Their red TADF emitter **TAT-FDBPZ** (**5**) showed an electroluminescence wavelength of 611 nm, a ΔE_{ST} of 0.10 eV, and a PLQY of 62%, but a low maximum EQE of 9.2%.[44]

Besides the dibenzo[a,c]phenazine-based TADF emitters, aromatic imide-based TADF molecules were also investigated as promising candidates for OLED emitters. Yang and coworkers reported an example of a 1,8-naphthalimide-based red TADF emitter.[45] **NAI-R3** (**6**), shown in Figure 3, depicted a narrow ΔE_{ST} of 0.059 eV, an electroluminescence wavelength of 622 nm, and a high maximum EQE of 22.5%. Recently, another orange-red-emitting, imide-based TADF emitter, ***P*-DMAC-BPyM** (**7**), was reported by You and coworkers, and is shown in Figure 3.[46] The novel acceptor molecule design of ***P*-DMAC-BPyM** (**7**) originates from the fusion of pyrazine and maleimide, which both show high electron deficiency, to induce a strong charge-transfer character. Furthermore, the highly twisted, rigidly locked structure effectively separates the frontier molecular orbitals, prevents non-radiative de-excitation, and inhibits quenching due to intermolecular π-π-stacking through its steric demand. The resulting TADF emitter exhibits an ΔE_{ST} of approximately zero, an electroluminescence wavelength of 579 nm, a PLQY of 72.2% in CBP-doped film, and an outstanding EQE_{max} of 26%.

In summary, the efforts of designing an orange-red TADF emitter for use in OLEDs hint toward overcoming the challenge of non-radiative de-excitation through rigid, electron-deficient aromatic systems with an intramolecular twist that separates the donor units from the acceptor core and ensures an adequate steric demand. Furthermore, the investigation of the usage of multiple donors shows that this strategy requires fine-tuning in dependence of the donors and acceptors. Lastly, substituents acting as auxiliary acceptors can be employed to fine-tune the LUMO level and red-shift emission.

1.2.2. TADF Fluorophores for Cellular Imaging

Thermally activated delayed fluorescence materials are increasingly explored for imaging purposes.[3, 8] While their benefits are manifold, the delayed fluorescence is especially interesting for time-resolved luminescence imaging (TRLI). While the lifetime of prompt fluorescence (10^{-10} to 10^{-7} s) can hardly be differentiated from the background signals during TRLI, the lifetime of delayed fluorescence ($>10^{-8}$ s) is sufficient.[8] Figure 4 shows the structures of TADF molecules successfully applied in TRLI.

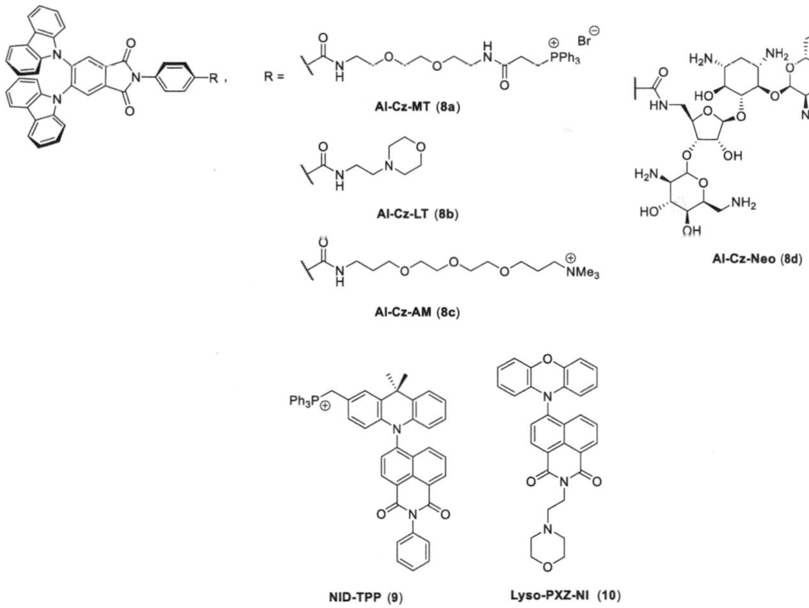

Figure 4. TADF emitters used in time-resolved luminescence imaging (TRLI).[47-50]

In 2019, Hu and coworkers reported upon **AI-Cz-MT** (**8a**) and **AI-Cz-LT** (**8b**), two TADF emitters that showed organelle specificity and their use in TRLI.[47] The molecule design included a mitochondria-targeting triphenylphosphonium (TPP) unit and a lysosome-targeting 2-morpholinoethylamine moiety, respectively. Both TADF molecules showed small ΔE_{ST} values of 0.12 eV and 0.11 eV and oxygen-affected delayed fluorescence in toluene, respectively. The emission wavelength of both emitters peaked at 520 nm. These emitters did not show cytotoxicity in MTS cell viability assays and were shown to enter live HepG 2 cells readily. Furthermore, the delayed fluorescence lifetimes in HepG 2 cells were 16.7 μs and 28.0 μs for **AI-Cz-MT** (**8a**) and **AI-Cz-LT** (**8b**), respectively. The short-lived autofluorescence could be circumvented and remarkably improved the signal-to-noise ratio. In their molecule design rationale, the well-known challenge of oxygen-quenching of TADF emitters was addressed by including the fluorophores in specific organelles, which was intended to lead to less exposure to oxygen.[47] Yang and coworkers also investigated this strategy with the focus on a different TADF core structure, namely 6-(9,9-dimethylacridin-10(9*H*)-yl)-2-phenyl-1*H*-benzo[*de*]isoquinoline- 1,3(2*H*)-dione (NID).[48] **NID-TPP** (**9**) featured a TPP unit and showed a minuscule ΔE_{ST} of 0.03 eV but only showed very weak emission and no TADF as a single molecule due to intramolecular rotation. Through the positive charge of the TPP unit, **NID-TPP** (**9**) accumulated into the mitochondrial matrix, where it aggregated. In the aggregated state, **NID-TPP** (**9**) showed aggregation-induced delayed fluorescence (AIDF) enhancement which resulted in stronger emission at $\lambda_{em} = 610$ nm, TADF, and an oxygen-independent long-lived fluorescence. The aggregation repressed intramolecular rotation and thus led to radiative de-excitation of the excited state. In addition to the TPP-conjugated **NID-TPP** (**9**), another NID-based emitter featuring a morpholine moiety, namely **Lyso-PXZ-NI** (**10**), was reported in 2020 by Yoon and coworkers.[49] In parallel to the approach taken by Hu and coworkers, the 2-morpholine unit was installed to target lysosomes. The ΔE_{ST} of **Lyso-PXZ-NI** (**10**) was 0.077 eV, with an emission wavelength of $\lambda_{em} = 614$ nm and a delayed fluorescence lifetime of 1.3 μs in PMMA thin films (10 wt%). This luminophore also showed AIDF as aggregation impeded energy transfer from the excited triplet state to surrounding oxygen molecules by which the non-radiative deactivation was inhibited and the radiative deactivation favored. Furthermore, MTT assay results showed very low cytotoxicity in living cells, and **Lyso-PXZ-NI** (**10**) displayed good cell permeability, organelle specificity, and long fluorescence lifetime signals in HeLa cells.[49]

Hu and coworkers extended the phthalimide-based TADF emitter series through **AI-Cz-AM** **(8c)**, an amphiphilic fluorophore monomer that consisted of the lipophilic D-A structure and the hydrophilic chain with a positive terminal charge, as depicted in Figure 4.[51] This monomer then spontaneously self-assembled into the nanoparticles **AI-Cz-NP** in an aqueous solution. The positive charge on the nanoparticle surface greatly aided cell penetration. Moreover, the aggregation benefited the photophysical performance as it restricted the intramolecular rotation, making the radiative de-excitation more favorable. It countered oxygen quenching of the excited triplet state through a more difficult oxygen diffusion to the D-A core. The emission wavelength of **AI-Cz-NP** was 550 nm with a PLQY of 1.36% in water, a low ΔE_{ST} (0.12 eV), and a long fluorescence lifetime of 15 µs in HepG 2 cells. In the same year, Hu and coworkers reported on the phthalimide-carbazole-based TADF emitter conjugated with neomycin, a bacterial 16S ribosomal RNA-targeted moiety called **AI-Cz-Neo** **(8d)** (Figure 4).[50] In PBS solution, it showed an emission wavelength maximum at 545 nm, and the ΔE_{ST} was calculated to a value of 0.22 eV. **AI-Cz-Neo** **(8d)** did not show cytotoxicity in the MTS cell viability assay and selectively labeled bacterial cells through uptake into those cells. No uptake of **AI-Cz-Neo** **(8d)** into mammalian cells was observed, in contrast to its analog without the neomycin entity. The selectivity of staining only bacterial and not mammalian cells was hypothesized to stem from membrane permeability issues. The fluorescence emission lifetime of a 1:1 complex of **AI-Cz-Neo (8d)** and an A-site of 16S rRNA in PBS was 17.8 µs. Furthermore, TRLI studies conducted in a coculture of Raw 264.7 macrophages and *S. aureus* or *E. coli* MG1655 displayed long average lifetime emissions of 28.7 µs and 26.7 µs, respectively.

In summary, the investigation of TADF luminophores suitable for TRLI showed that the challenge of de-excitation of the excited triplet state through oxygen quenching could either be impeded by (i) organelle specificity using terminal groups like TPP or 2-morpholine units and (ii) through aggregation or directed assembly into nanoparticles, which both reduce the exposure of the TADF core to oxygen. Furthermore, depending on conjugated moieties, such as neomycin, different cell types can be specifically stained.

1.2.3. TADF Emitters as MOF Linker Materials

In 2018, Adachi and coworkers reported the first metal-organic framework (MOF) exhibiting intrinsic thermally activated delayed fluorescence.[52] MOFs combine metal-containing nodes and organic linkers to establish hybrid porous materials.[53] Due to their defined, rigidified structure, non-radiative deactivation is reduced, and a higher photoluminescence quantum yield was observed.[54-55] Furthermore, the structural rigidity was preventive against aggregation-caused quenching effects, which further benefits the efficiency of the luminophores used as linkers in MOFs.[56] A schematic structure of the **UiO-68-dpa** (**11**) Zr-based MOF and its **H₂tpdc-dpa** (**12**) TADF linker that the Adachi group investigated are shown in Figure 5.[52] **H₂tpdc-dpa** (**12**) is composed of electron-donating and -accepting units separated through a twisted structure induced by a large dihedral angle. This molecule design led to a small experimentally obtained ΔE_{ST} value of 0.24 eV. Its emission wavelength was λ_{max} = 461 nm (2 wt% in PMMA), and the lifetime of the delayed fluorescence was τ_d = 199 ms. The investigation of the TADF linker at different doping concentrations in PMMA thin film showed concentration-caused quenching, which resulted in reduced delayed emission lifetimes. Furthermore, the photoluminescence quantum yield was 39% under a nitrogen atmosphere. On the other hand, **UiO-68-dpa** (**11**) showed green emission with λ_{max} = 501 nm. The red-shift of the emission and a decreased delayed emission lifetime (τ_d = 0.18 ms) were attributed to the coordination interaction between the Zr metal ion and carboxylic acid moiety as well as the interaction between two linkers in close proximity within the MOF structure.[52] The ϕ under nitrogen atmosphere was 30%. In contrast to the linker, the Zr-based MOF showed a more severe depression of photoluminescence quantum yield when exposed to air, and organic solvents as the PLQY dropped to 18%. This was ascribed to the observation that the delayed component was completely suppressed by oxygen-caused triplet quenching.

Wöll and coworkers presented the TADF surface-anchored MOF (SURMOF) thin-film **Zn-DPA-TPE** (**15**) in 2020.[55] The linker **DPA-TPA** (**16**) does not show fluorescence in solution because of the free rotation at the tetraphenylethylene moiety that leads to non-radiative deactivation of the excited state. However, by enforced, dense packing in a highly ordered structure as given in a SURMOF, based on Zn-paddlewheel secondary building units (SBUs) that are connected through the linkers, the radiative de-excitation is enabled, and TADF with a delayed emission lifetime of τ_d = 200 μs was observed. The quantum yield was determined as ϕ = 14 ± 2%. TD-DFT calculations for the **Zn-DPA-TPE** (**15**) MOF indicated a ΔE_{ST} of roughly 0.24 eV. Furthermore, Wöll and coworkers reported upon the implantation of their TADF MOF in an LED device.

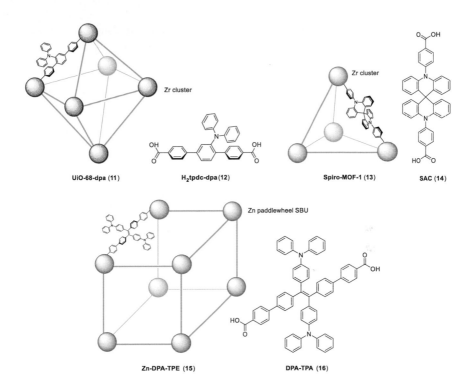

Figure 5. TADF emitters as linkers in MOFs as reported by Adachi and coworkers,[52] Yang and coworkers,[57] and Wöll and coworkers.[55]

Yang and coworkers recently developed the TADF MOF **Spiro-MOF-1 (13)** formed by Zr-clusters linked by the organic linker unit **SAC (14)**.[57] The molecule design of **SAC (14)** originated from the desire to have a highly twisted and locked structure to minimize single bond rotation, which was realized by utilizing a *spiro*-C-atom. While the electron-accepting moiety was localized on the benzoic acid moieties through DFT calculation, the electron-donating units were located on the *spiro*-bi-acridine. **SAC (14)** showed an emission wavelength maximum at 484 nm in degassed THF solution with a ϕ of 25% and a delayed emission lifetime of $\tau_d = 1.18$ μs. Upon exposure to air, the delayed lifetime decreased sharply to $\tau_d = 0.38$ μs, showing oxygen-induced quenching of the fluorescence of the organic linker. **Spiro-MOF-1 (13)** based on **SAC (14)**, displayed a narrow ΔE_{ST} of approximately 0.14 eV. The photophysical investigation of bulk **Spiro-MOF-1 (13)** showed that the emission wavelength maximum und air was $\lambda_{max} = 521$ nm, while the delayed lifetime was $\tau_d = 712$ ns and the photoluminescence quantum yield ϕ = 8.6%. **Spiro-MOF-1 (13)** was subsequently drafted into nanoparticles (**nMOF**) and further modified by polyethylene glycol to obtain **nMOF-PEG**, a MOF designed for improved biocompatibility. Both **nMOF** and **nMOF-PEG** depicted lower cytotoxicity than **SAC (14)** or **4CzIPN**, even at higher concentrations. **nMOF-PEG** in cells showed an emission wavelength maximum at 512 nm, a ϕ of 7.3%, and a delayed emission lifetime of τ_d = 204 ns.[57]

In summary, embedding TADF linkers into highly ordered MOF structures aids in avoiding non-radiative de-excitation and losses. However, oxygen quenching remains a challenge.

1.3. Menthol and Other TRPM8 Agonists and Antagonists

In 2021, the Nobel prize in physiology or medicine was awarded to David Julius and Ardem Patapoutian for their groundbreaking efforts in discovering of receptors for temperature and touch.[58] Both researchers independently identified the transient receptor potential melastatin subtype 8 (TRPM8), which is activated by cold temperatures (< approx. 25-28 °C) or cooling agents such as menthol.[59-61] TRPM8 is a Ca^{2+}-permeable cation channel, albeit being non-selective towards monovalent and divalent cations.[62] Furthermore, TRPM8 was highly expressed in several tumor cells, such as prostate cancer cells, and its overexpression induces cold hypersensitivity, which is linked to various inflammatory and (neuropathic) pain states and other discomfort states.[63-64] Menthol and its derivatives were found to act as agonists and antagonists for TRPM8 and find use in food, pharmaceutical, and cosmetic industries.[61] Some examples of menthol-based TRPM8 agonists and an antagonist are shown in Figure 6.

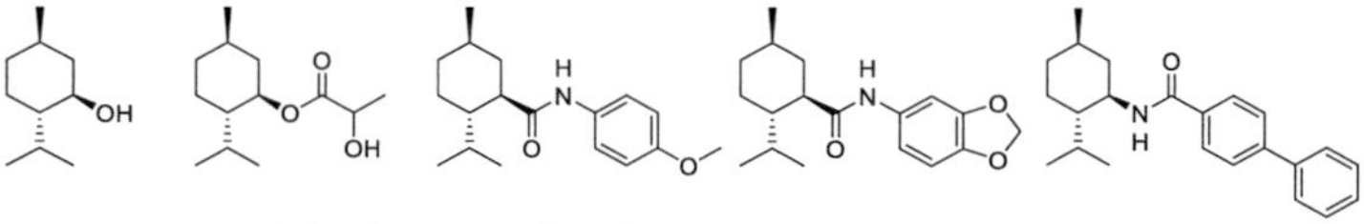

(–)-menthol (17) Frescolat ML (18)   WS-12 (19)   20   (-)-menthyl 1 (21)

Figure 6. Agonists and an antagonist for TRPM8 ion channels.

(–)-Menthol (17) is well-known for its topical cooling effect.[65] While menthol has been used for centuries in various cultures,[66] understanding the mechanism of ligand-dependent TRPM8 channel gating is still under investigation.[67-69] Nilius and coworkers reported in 2004 and 2005 that menthol induces TRPM8 channel activation by shifting the voltage dependence of activation. Thereby, the channel opens at physiologically relevant voltages.[70-71] Furthermore, Tsuzuki et al. reported in 2004 that menthol induces Ca^{2+} release from presynaptic, intracellular Ca^{2+} stores of sensory neurons in a Ca^{2+}-imaging study on dorsal root ganglion (DRG).[72] The naturally occurring menthol derivative **Frescolat ML (18)** (Figure 6) was found to exhibit a half-maximal effective concentration (EC50) of 163 ± 49 µM in *Xenopus* oocytes; in comparison to **(–)-menthol (17)**, which displayed an EC50 $= 196 \pm 22$ µM.[61, 73] Based on menthol, more potent TRPM8 agonists such as the carboxamide **WS-12 (19)** were developed. **WS-12 (19)** activated TRPM8 channels expressed in *Xenopus* oocytes at an EC50 $= 12 \pm 5$ µM.[74] The carboxamide **20** featuring a 1,3-benzodioxole unit was reported to show a cooling intensity 100 times stronger than menthol in a patent by Givaudan SA.[75] Moreover, in 2021, Journigan et al.[76] reported upon the TRPM8 antagonist **(–)-menthyl 1 (21)**, a structure previously reported in 2010 by Ortar et al.[77] In a human TRPM8 (hTRPM8) orthologue, **(–)-menthyl 1 (21)** was found to inhibit a menthol-evoked Ca^{2+} response with a half-maximal inhibitory concentration (IC50) value of 805 ± 200 nM while for the rat TRPM8 orthologue (rTRPM8), it was more efficacious with an IC50 $= 117 \pm 18$ nM.[76]

1.4. Azobenzene-Based Photoswitches

Azobenzenes (ABs) are a class of compounds known for over 150 years.[78] In the 1930s, the isomerization from the *trans*- to the *cis*-azobenzene **22** was discovered.[79] The *trans*-azobenzene is more stable than its *cis*-isomer by 10-12 kcal mol^{-1} and can be isomerized upon irradiation with light into the *cis*-azobenzene.[80] The reversible photoisomerization is a characteristic trait of photoswitches. The isomerization of azobenzene leads to conformational changes, a change in polarity, and a difference in the end-to-end distance.[80] The reverse *cis*→*trans*-isomerization of ABs can be induced photochemically or thermally (Scheme 1).

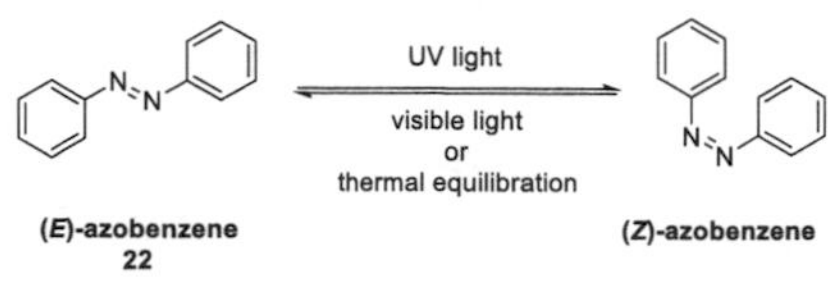

Scheme 1. Photoisomerization of azobenzene **22**.

Azobenzenes demonstrate interesting photophysical behavior. The excitation of the *trans*-azobenzene from the ground state S_0 to S_1 is a weak $n \rightarrow \pi^*$ transition, observed at a wavelength of approx. 450 nm, while the absorption of energy leading to the excitation from S_0 to S_2 ($\pi \rightarrow \pi^*$) is stronger and observed at roughly 300 nm. As a consequence of the isomerization of the N=N bond, the metastable *cis*-azobenzene is formed. The isomerization to the *cis*-isomer is mirrored in the absorption spectra, as a slight shift and an intensification of the $n \rightarrow \pi^*$ transition is observed, whereas the intensity of the $\pi \rightarrow \pi^*$ transition is reduced and hypsochromically shifted. The difference in the absorption spectra of the light-induced interconversion of the two azobenzene isomers indicates photochromism. Furthermore, azobenzene does not adhere to Kasha's rule (*vide ante*) as the isomerization can occur from either the S_1 or S_2 state.[78] The mechanism and underlying photophysics of the isomerization upon energy uptake followed by de-excitation is a matter which was and still is subject to discussion.[81-87]

Thermal relaxation from *cis→trans* takes days at room temperature.[88] However, depending on the application of an AB-based photoswitch, a faster reverse isomerization might be desirable. Furthermore, excitation at longer wavelengths is preferred for the use in biological systems.[89] Both requirements can be addressed by substituting azobenzenes. For example, installing electron-donating and electron-accepting groups on either side in *para*-position leads to the formation of "push-pull"-systems that show red-shifted absorption. Moreover, red-shifted azobenzenes were found to show faster thermal relaxation rates.[90]

The principle behind using photoswitches in photopharmacological applications is well depicted in the work by Kramer, Trauner, and coworkers, who reported the AB-based photoswitch **MAL-AZO-QA (23)** shown in Scheme 2 for the use in synthetic photoisomerizable azobenzene-regulated K^+ (SPARK) channels.[91-92] The rationale behind the molecule design was to install a quaternary amine (QA) unit, which blocks the pore of K^+ channels, and a cysteine-reactive maleimide (MAL) group to a photoswitchable azobenzene. The tetraethylammonium moiety in the *trans*-**MAL-AZO-QA (23)** blocked the voltage-gated Shaker K^+ channel and thereby impeded the ion flow. Upon irradiation with UV light at $\lambda = 380\text{-}390$ nm, the *trans*-isomer was switched to the *cis*-isomer, which considerably shortened the end-to-end distance from approx. 17 Å to 10 Å. A difference of several Å was reportely decisive for an ineffective vs. effective block,[93] and in accordance to this finding, the *cis*-**MAL-AZO-QA (23)** with an end-to-end distance of roughly 10 Å could not block the pore, and the K^+ current was turned on. Irradiation with λ = 500-505 nm accelerated the closure of the ion channel, and the original membrane potential was restored. This was the first example of an optical tool that effectively silenced a neuronal population.[94]

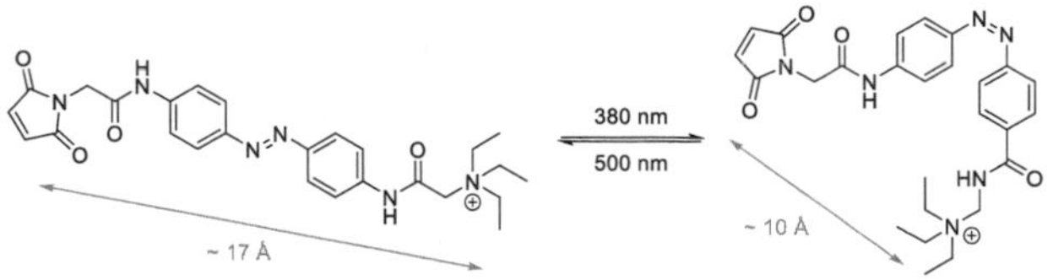

MAL-AZO-QA (23)

Scheme 2. Azobenzene-based photoswitch **MAL-AZO-QA** (**23**) for synthetic photoisomerizable azobenzene-regulated K^+ (SPARK) channels, as reported by Kramer, Trauner, and coworkers.[91-92]

Moreover, Trauner and coworkers investigated several AB-based photoswitches, which acted as agonists and antagonists to the vanillin receptor 1 (TRPV1), displayed in Figure 7. While menthol is the best-known agonist for TRPM8, capsaicin holds this role for TRPV1. TRPV1 is also a prominent member of the transient receptor protein (TRP) family and represents a non-selective cation channel that is permeable to Ca^{2+} ions while showing no bias regarding mono- or divalent cations.[95] While TRPM8 is a cold receptor, TRPV1 responds to temperatures above 43 °C amongst other chemical and physical stimuli and induces a sensation of pain and burning upon activation.[95-96]

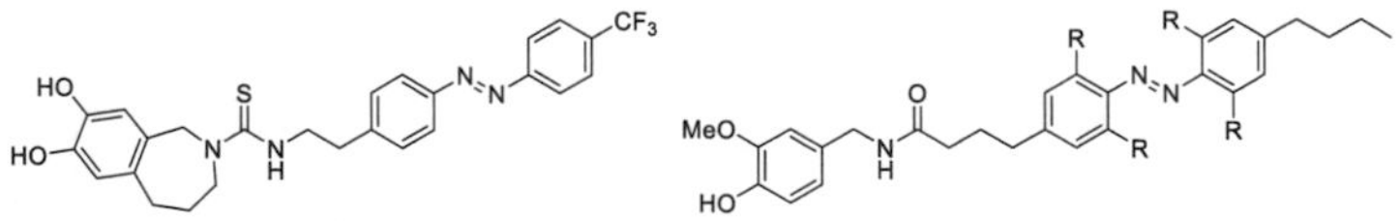

AC4 (24)

R = H (**AzCA4 (25a)**)
Cl (***red*-AzCA-4 (25b)**)

Figure 7. Examples of a TRPV1 antagonist **AC4** (**24**) and agonists **AzCA4** (**25a**) and ***red*-AzCA-4** (**25b**) as reported by Trauner and coworkers.[95-97]

In 2013, Trauner and coworkers introduced a photoswitch called **AC4 (24),** shown in Figure 7. The *trans*-isomer of **AC4** (**24**) could be isomerized upon irradiation at 360 nm, while the isomerization was reverted upon irradiation with a light of 440 nm. The investigation of **AC4** (**24**) showed that it acted as an antagonist of TRPV1 in both *trans*- and *cis*-form. Its *trans*-isomer operated as an antagonist of capsaicin-induced TRPV1 currents, while the *cis*-isomer was reported to be an antagonist of TRPV1 upon voltage-induced activation. The IC_{50} of **AC4**

(24) on capsaicin-induced TRPV1 currents in a Ca^{2+} luminescence essay was 3.1 ± 0.6 μM.[95] By applying the photoswitch AC4 (24), the TRPV1 ion channel was effectively converted into a photoreceptor. Trauner and coworkers continued their studies on the TRPV1 channels and reported on a TRPV1 agonist that enabled optical control over this ion channel without a second factor such as capsaicin in 2015. The TRPV1 agonist AzCA4 (25a) is displayed in Figure 7 and could be successfully switched from its *trans*-isomer to the *cis*-isomer with $\lambda = 365$ nm. The *cis-trans*-isomerization took place upon irradiation with $\lambda = 460$ nm. While AzCA4 (25a) was reported to show high efficacies in the *cis*-configuration, the agonist was rendered inactive in the dark at an optimized concentration. AzCA4 (25a) enabled the reversible optical control of DRG neuronal activity.[96] However, for the application in biological systems, short-wavelength light of high energy can be harmful, and long-wavelength light within the optical window (650-950 nm) is advantageous as it penetrates deeper into the tissue.[98] Thus, Trauner, Konrad, and coworkers developed red-shifted AB-based photoswitches by late-stage C-H chlorination and applied their synthetic methodology to AzCA4 (25a). The tetrachlorinated analog red-AzCA-4 (25b) could be switched from the *trans*- to the *cis*-form by irradiation with $\lambda = 560$ nm. The photoswitching was reversed by irradiation with $\lambda = 400$ nm. In accordance with the findings for AzCA4 (25a), the red-shifted red-AzCA-4 (25b) displayed a higher efficacy in its *cis*-form when operating on the TRPV1 channel.[97]

Beyond the use of ABs as photoswitches tethered to ligands or in photochromic ligands, they have also been investigated as fluorophores. Although the usage of ABs as fluorophores might not be entirely intuitive as the absorbed energy is already put to use in the photoisomerization, investigations of AB-based fluorophores have been made. In 2012, Tamaoki and coworkers reported an azobenzene-based structure for which fluorescence could be switched on upon photoisomerization from 1Ct (26a) to 1Oc (26b), as shown in Scheme 3.[99] They reported a water-soluble and photoswitchable rhodamine spiroamide (RSA) derivative, which underwent ring-opening upon *trans*→*cis* photoisomerization with UV light ($\lambda = 365$ nm) and became fluorescent. The isomerization was reverted by irradiating the material with $\lambda = 490$ nm or through thermal relaxation. The fluorescence quantum yield for the photostationary state (PSS) in an aqueous solution at pH 3.8 was determined to be 0.37.

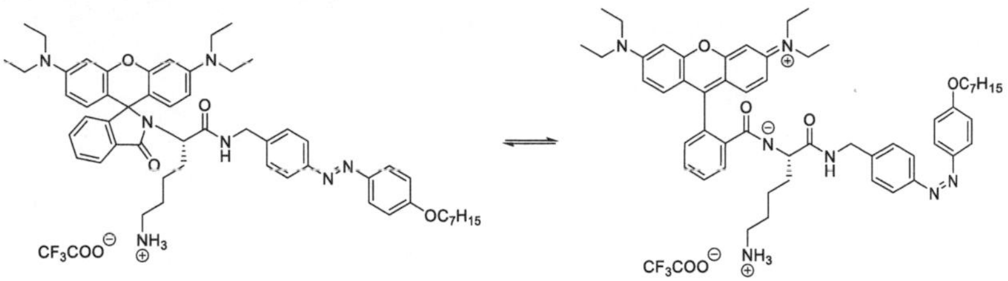

1Ct (**26a**)
non-fluorescent

1Oc (**26b**)
fluorescent

Scheme 3. Azobenzene with photoswitchable on-off fluorescence as reported by Tamaoki and coworkers.[99]

Other approaches to fluorescent ABs were based on the formation of aggregates or structures of higher order.[100-103]

2. Objective

This work was focused on molecule design and organic synthesis, and it can be subcategorized as follows: (i) development of novel acceptors for red thermally activated delayed fluorescence (TADF) emitters, (ii) synthesis of photoswitchable menthol derivatives, and (iii) development of photoswitchable push-pull azobenzene systems.

Red emitters and organic light-emitting diodes (OLEDs) find broad applications in bioimaging, sensors, telecommunication, and night vision.[36] Amongst the electron-accepting systems for TADF molecules that were investigated, dibenzo[a,c]phenazine (BP) **I** (Figure 8) is identified as a promising candidate, and it was first mentioned in the context of TADF in 2018.[42-44, 104-113] The aim of the BP-related studies is to investigate (poly-)donor-acceptor systems and a synthesis route leading to easy synthetic access to substituted BP acceptors. The applications for which these molecules are designed to vary from emitters in OLEDs, linkers for MOFs, and fluorophores for insertion into MOF cavities and cellular imaging purposes.

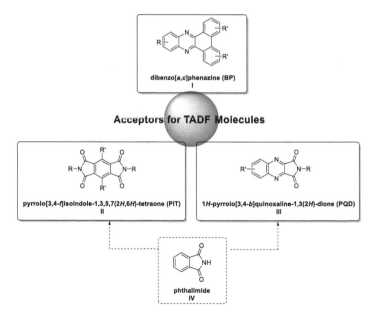

Figure 8. Development of novel acceptors for red TADF emitters.

36

Furthermore, the designs of aromatic imide-based pyrrolo[3,4-*f*]isoindole-1,3,5,7(2*H*,6*H*)-tetraone (PIT) **II**- and 1*H*-Pyrrolo[3,4-*b*]quinoxaline-1,3(2*H*)-dione (PQD) **III**-based acceptors are established (Figure 8). Both molecule designs originate from the motivation to further develop the phthalimide **IV**-based TADF emitters investigated in the Bräse group previously.[114] The enlargement of the acceptor system and the addition of more heteroatoms are intended, as visualized in Figure 8. Both measures are targeted at red-shifting the emission leading to fluorophores emitting longer wavelengths. After the synthesis, the emitters are characterized by cyclic voltammetry, UV-vis and fluorescence spectroscopy, and standard characterization methods for organic compounds.

Photoexcitation results in an excited state from which de-excitation leads to luminescence – a process desired for (TADF) fluorophores. However, excited states can also be used for other purposes, such as the photoisomerization of photoswitches. Photoswitches experience a change of configuration upon absorption of energy through irradiation. This reversible configurational change is an extremely valuable feature as it enables the on- and off-switching of a molecular structure. The optical control that is thereby obtained can be applied in photopharmacology.[115-116] Azobenzenes (ABs) **V** are one of the most prominent and well-studied classes of photoswitches. AB can serve as the photoisomerizable unit attached to receptor ligands that act as agonists or antagonists. In this work, (−)-menthol (**17**)-derived molecules containing an azobenzene structural unit were synthesized for the application in TRPM8 channel investigations (Figure 9).

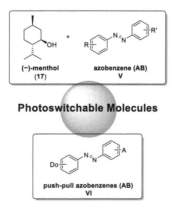

Figure 9. Synthesis aim regarding photoswitchable molecules targeting TRPM8 ion channels and push-pull systems **VI** carrying donor (Do) and acceptor (A) units.

Lastly, both concepts – thermally activated delayed fluorescence and photoisomerization of azobenzenes – were combined to investigate push-pull structures **VI** that would allow an electron-donating and -accepting unit to come into spatial proximity upon photoisomerization (Figure 9).

3. Results and Discussion

3.1. Dibenzo[a,c]phenazine (BP)-Based TADF Molecule Design

The study of novel TADF emitters was started with molecule designs based on the dibenzo[a,c]phenazine (BP) acceptor due to its previously investigated suitability for orange-red OLEDs as described in chapter 1.2.1. The investigations regarding the BP acceptor can be sectioned into the following donor (D)-acceptor (A) conjugate types: (i) D-A (**VII**), (ii) 2D-A-D (**VIII**), and (iii) 2D-A-R (**IX**), as visualized in Figure 10.

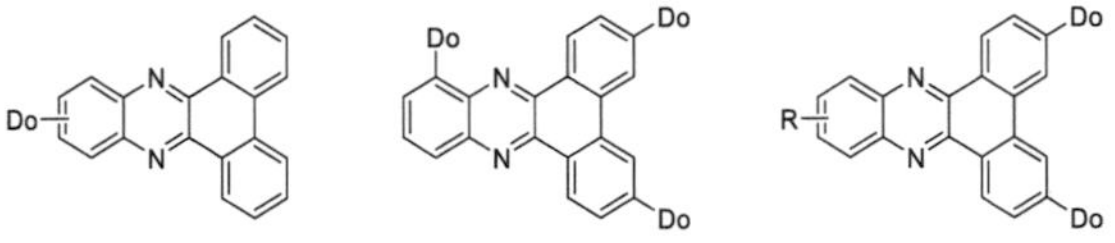

Figure 10. Dibenzo[a,c]phenazine-based donor-acceptor conjugate types **VII-IX**, Do = donor.

3.1.1. Dibenzo[a,c]phenazine-Based D-A-Type Structures

D-BP-Type Structures – Synthesis

Firstly, the basic D-A system was investigated. The dibenzo[a,c]phenazines **29a** and **29b** were prepared according to a literature-reported procedure[107] as shown in Scheme 4, with yields of 27 and 96% as yellow solids, respectively. The synthesis of compound **29b** was also recently described by Kwon and coworkers, as well as Yang and coworkers and Zhou et al.[107, 117-118]

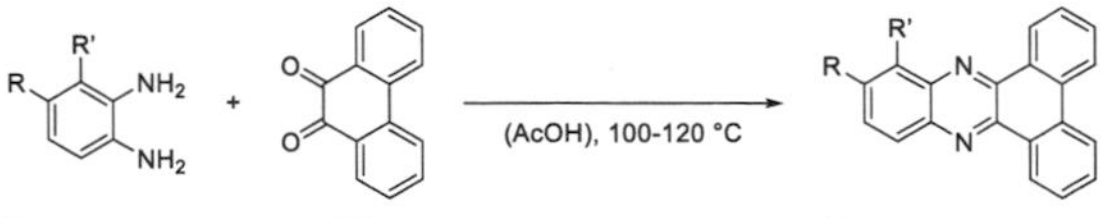

27a R = H, R' = F   28a
27b R = Br, R' = H

29a R = H, R' = F (27%)
29b R = Br, R' = H (96%)

Scheme 4. Preparation of the dibenzo[a,c]phenazine acceptor **29a** and **29b**.

Subsequently, donor units were attached. In the case of 10-fluorodibenzo[a,c]phenazine **29a**, nucleophilic aromatic substitution was used to yield D-A conjugates **PXZ-BP** (**30a**),

DMAC-BP (30b) and **DTCz-BP (30c)** with yields of 39, 30, and 68% as a red, orange and yellow solid, respectively, as depicted in Scheme 5.

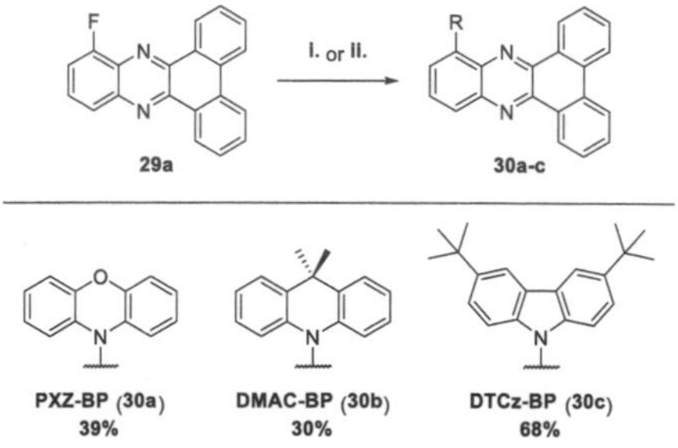

Scheme 5. Synthesis of D-A-type BP-based molecules *via* nucleophilic aromatic substitution; (i) 3,6-di-*tert*-butyl carbazole, K_3PO_4, (DMSO), 100 °C, (ii) 10*H*-phenoxazine or 9,9-dimethyl-9,10-dihydroacridine, NaH, (DMF), 100 °C.

PXZ-BP (30a) and **DMAC-BP (30b)** shown in Scheme 5 were published by Yang and coworkers.[107] They obtained superior yields by using 10-bromodibenzo[*a,c*]phenazine in a BUCHWALD-HARTWIG amination reaction. **DTCz-BP (30c)** is not literature-reported to the best of my knowledge (Scifinder[n] structure search on 05/03/2022), but a similar compound with two DTCz units was patented in 2017 by Cao *et al.*[119]

The 11-bromodibenzo[*a,c*]phenazine **29b** acceptor was modified to obtain a D-A conjugate *via* BUCHWALD-HARTWIG cross-coupling with a DTCz donor **31c** shown in Scheme 6. Analogous structures employing a DMAC and PXZ donor were reported.[107, 117] Lee and coworkers reported a similar structure with two other nitrile groups attached to the acceptor core.[106]

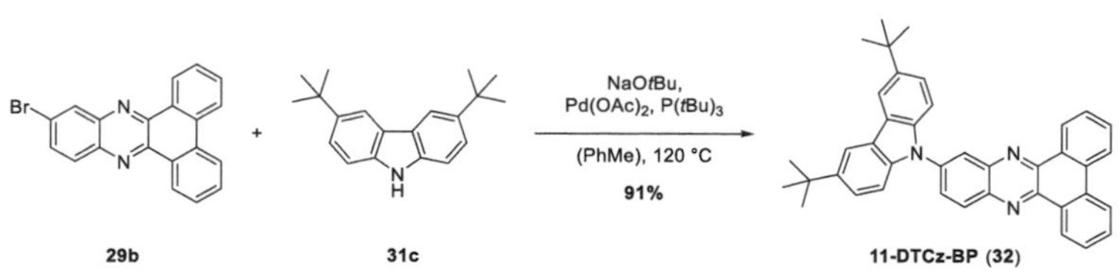

Scheme 6. Synthesis of D-A **11-DTCz-BP (32)**.

After the synthesis, **11-DTCz-BP (32)** was tested in a MOF structure in collaboration with Knebel and coworkers. The molecule design of **11-DTCz-BP (32)** had been customized to fit the pocket of the MIL-68(In) MOF, and Bahram Hosseini Monjezi carried out the experiments regarding the incorporation of the dye into the MOF. The photophysical measurements were conducted in a corporation with further collaboration partners. The first results indicate that the fluorophore shows TADF properties and could successfully be embedded into the MIL-69(In) pores. Furthermore, the photoluminescence quantum yield of the TADF-MOF system was increased compared to the single TADF molecule.

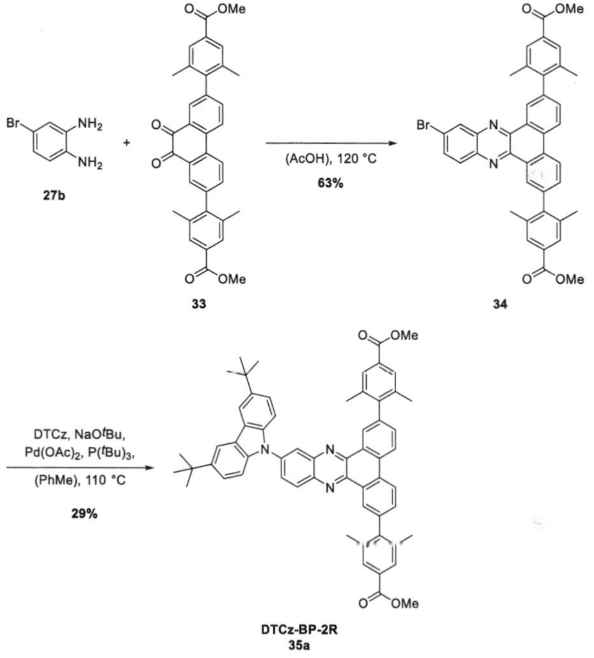

Scheme 7. Synthesis of **DTCz-BP-2R** (**35a**), where the starting material **33** was provided by Simon Oßwald (Bräse group).

The application of TADF in MOF structures is highly promising as discussed in chapter 1.2.3, and can also be implemented through the design of TADF-linkers for MOFs. Therefore, a BP-based D-A conjugate **DTCz-BP-2R** (**35a**) (Scheme 7) with an elongated π-backbone was designed. To coordinate the metal centers, terminal ester groups implemented in the backbone can be easily hydrolyzed in the final step.

Since solubility issues occurred in the initial design without additional methyl groups, these substituents were introduced to increase solubility and ensure a high backbone twist. The synthesis was realized as depicted in Scheme 7. In the first step, the starting material **33**, which was kindly provided by Simon Oßwald (Bräse group), was used in a condensation reaction in acetic acid at 120 °C to obtain intermediate **34** in a yield of 63% as a pale-yellow solid. Subsequently, it was employed in a BUCHWALD-HARTWIG amination step to yield **DTCz-BP-2R** (**35a**) in 29% yield as a bright yellow solid.

The D-A-type compounds based on a dibenzo[*a,c*]phenazine acceptor core were investigated. Three different electron-donating units, namely DTCz, DMAC, and PXZ, were used and yielded **PXZ-BP** (**30a**), **DMAC-BP** (**30b**) and **DTCz-BP** (**30c**) in moderate to good yields. Furthermore, BP-based fluorophores for the application in metal-organic frameworks were designed and prepared. **11-DTCz-BP** (**32**) was customized to fit the size requirements of the MIL-68(In) MOF and is currently under further investigation in cooperation with collaboration partners. Lastly, the D-A-type BP-based linker compound **35a** was prepared and will be further investigated in collaboration with the Wöll group.

D-BP-Type Structures – DFT Calculations

In addition to the synthesis, the BP-based D-A conjugates were investigated utilizing TD-DFT calculation. The results in this subchapter were provided by Changfeng Si (Zysman-Colman Group, University of St. Andrews). The computations of the geometry-optimized ground states were conducted in the gas phase using the Pbe1pbe/6-31G(d,p) functional, and the excited states were computed using time-dependent DFT within the TAMM-DANCOFF approximation (TDA) based on the optimized ground-state geometries.

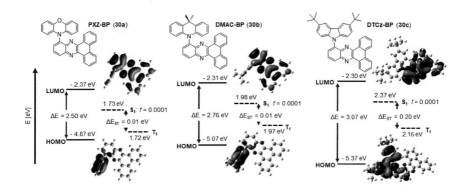

Figure 11. Geometry optimization and excited state (TD)-DFT calculation results for D-A-type structures **PXZ-BP (30a)**, **DMAC-BP (30b)**, and **DTCz-BP (30c)** provided by Changfeng Si (Zysman-Colman Group, University of St. Andrews); computation in the gas phase using the Pbe1pbe/6-31G(d,p) functional.

When comparing the DFT-calculated geometry-optimized ground states of **PXZ-BP (30a)**, **DMAC-BP (30b)**, and **DTCz-BP (30c)** which are depicted in Figure 11, the LUMOs are located on the dibenzo[a,c]phenazine moiety and are found to be in a similar range as the values amount to -2.37, -2.31, and -2.30 eV, respectively. The HOMOs, however, are mainly distributed on the electron-donating units. For **DTCz-BP (30c)**, the calculation result shows that the HOMO is slightly spread out further onto the BP acceptor in comparison to **PXZ-BP (30a)** and **DMAC-BP (30b)**. This is likely due to a smaller dihedral angle between the DTCz donor and BP acceptor leading to a worse separation of the frontier molecular orbitals. Confirming the donor strength, the HOMO levels of **PXZ-BP (30a)**, **DMAC-BP (30b)** and **DTCz-BP (30c)** are found at -4.87, -5.07, and -5.37 eV, respectively.[36] Thereby, the HOMO-LUMO energy gap of **PXZ-BP (30a)**, **DMAC-BP (30b)**, and **DTCz-BP (30c)** amount to 2.50, 2.76, and 3.07 eV, respectively.

The excited states were computed using TD-DFT calculations. The S_1 and T_1 state are almost exclusively charge-transfer transitions between the HOMO and LUMO for **PXZ-BP (30a)** and **DMAC-BP (30b)** (HOMO→LUMO, 99%). While the nature of the S_1 state of DTCz-BP is also calculated to be a 99% CT transition, the T_1 state exhibits an 86% CT transition character. The computational values of S_1 and T_1 result in a ΔE_{ST} of 0.01 eV for both **PXZ-BP (30a)** and **DMAC-BP (30b)**, with indicates a very narrow energy gap that leads to an efficient upconversion of triplet excitons. In contrast, the computed ΔE_{ST} value for **DTCz-BP (30c)** amounts to 0.20 eV. This is likely due to the increased overlap of the HOMO and LUMO, which indicates a less effective separation of the FMOs. Lastly, the oscillator strengths corresponding to an S_0-S_1 transition amount to 0.0001 for all three emitters shown in Figure 11.

In comparison to these results, Yang and coworkers had performed the theoretical calculations using a Gaussian 09 package at the level of B3LYP/6-31G(d,p).[107] Their results for the ground state geometry optimization indicated a large dihedral angle of 85° for **PXZ-BP (30a)** and **DMAC-BP (30b)**, which led to a good separation of the FMOs and small ΔE_{ST}. They also reported a minuscule ΔE_{ST} of 0.01 eV for both compounds, calculated using TD-DFT calculations using a PBE0/def2-SVP functional.

Furthermore, TD-DFT calculations were carried out for **11-DTCz-BP (32)**, an isomeric compound to **DTCz-BP (30c)**. This compound had been designed for the loading into MOF pores and the results of the ground-state geometry optimization and excited states calculation are shown in Figure 12. Similar to the observation made for **DTCz-BP (30c)**, **11-DTCz-BP (32)** shows the LUMO mainly on the BP core, while the HOMO is mainly located on the carbazole-based donor. The FMOs also show an increased overlap compared to **PXZ-BP (30a)** and **DMAC-BP (30b)**. The HOMO energy level was calculated to be -5.48 eV, while the LUMO energy was determined as -2.25 eV. Consequently, the HOMO-LUMO gap amounts to 3.23 eV. The excited state simulation shows that the S_1 and T_1 energies amount to 2.68 and 2.29 eV, respectively, which results in a ΔE_{ST} of 0.38 eV. Furthermore, the oscillator strength was calculated to be 0.25.

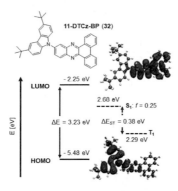

Figure 12. Geometry optimization and excited state (TD)-DFT calculation results for **11-DTCz-BP (32)** provided by Changfeng Si (Zysman-Colman Group, University of St. Andrews); computation in the gas phase using the Pbe1pbe/6-31G(d,p) functional.

In comparison to **DTCz-BP (30c)**, the LUMO level of **11-DTCz-BP (32)** is destabilized by 0.05 eV, while the HOMO level is deepened by 0.11 eV. The HOMO-LUMO gap is increased for **11-DTCz-BP (32)** ($\Delta E_{HOMO\text{-}LUMO} = 3.23$ eV), by 0.16 eV in comparison to **DTCz-BP (30c)** ($\Delta E_{HOMO\text{-}LUMO} = 3.07$ eV). This follows Yang and coworkers's findings on the DMAC- and PXZ-substituted analogs. They argue that the emitters substituted in the 10-positions have a shorter D-A centroid distance and a smaller orbital overlap indicating their stronger charge-transfer (CT) character and leading to narrowed HOMO-LUMO gaps.[107] Moreover, the TD-DFT calculation results show that **DTCz-BP (30c)** ($\Delta E_{ST} = 0.20$ eV) displays a lower ΔE_{ST} than **11-DTCz-BP (32)** ($\Delta E_{ST} = 0.38$ eV). Nonetheless, both emitters display a ΔE_{ST} value above 0.10 eV and are prone to depict less efficient RISC. When comparing the oscillator strengths, a significant increase is observed for **11-DTCz-BP (32)**, which hints at the possibility of higher photoluminescence quantum yields and radiative rates.[107]

In summary, the D-A-type compounds **PXZ-BP (30a)**, **DMAC-BP (30b)** and **DTCz-BP (30c)** were investigated employing (TD)-DFT calculation and showed an increasingly shallow HOMO with higher donor strength. Furthermore, the predicted ΔE_{ST} for the first two compounds is very low, resulting from well-separated FMOs, while the calculated ΔE_{ST} of **DTCz-BP (30c)** is larger. The isomer to the latter, **11-DTCz-BP (32)**, depicted a less shallow HOMO and an increased LUMO level, and a higher ΔE_{ST}, but an increased oscillator strength in comparison to **DTCz-BP (30c)**.

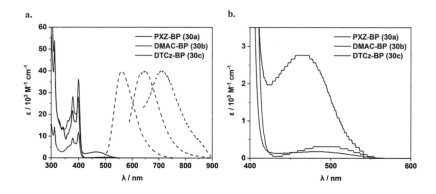

46

D-BP-Type Structures – Absorption and Emission

After the synthesis and theoretical calculations, D-A-based fluorophores **PXZ-BP** (**30a**), **DMAC-BP** (**30b**), **DCTz-BP** (**30c**), and **11-DTCz-BP** (**32**) were investigated employing UV-vis and fluorescence spectroscopy. The absorbance spectra of **DTCz-BP** (**30c**), **DMAC-BP** (**30b**), and **PXZ-BP** (**30a**) were measured by Changfeng Si (Zysman-Colman Group, University of St. Andrews) and are displayed in Figure 13.

Figure 13. UV-vis and fluorescence spectra for the D-A-type emitter series **PXZ-BP** (**30a**), **DMAC-BP** (**30b**), and **DTCz-BP** (**30c**) as provided by Changfeng Si (Zysman-Colman Group, University of St. Andrews), emission measurements in 0.1 mM toluene solutions are shown in dotted lines.

Figure 13 depicts the molar extinction coefficient ε against the wavelength. All three compounds show similar absorption profiles below 400 nm with prominent peaks at 378 nm and 399 nm, which can be attributed to locally excited (LE) vibronic absorption from the rigid BP acceptor, according to Yang and coworkers.[107] **DTCz-BP** (**30c**) shows another signal at 343 nm. While the **PXZ-BP** (**30a**) and **DMAC-BP** (**30b**) only show very low absorbance in the region above 400 nm (Figure 13b) with extinction coefficients of 0.31 and 0.17·10^3 M^{-1} s^{-1}, respectively, **DTCz-BP** (**30c**) shows an extinction coefficient of 2.76·10^3 M^{-1} s^{-1} for the absorption band at 468 nm. The low extinction coefficients for **PXZ-BP** (**30a**) and **DMAC-BP** (**30b**) follow the beforementioned findings of Yang and coworkers, who ascribe this phenomenon to a forbidden transition of their enhanced CT states.[107] In the case of **DTCz-BP** (**30c**), the absorption at 468 nm originates from an ICT transition.[106]

Furthermore, the emission wavelengths of **PXZ-BP** (**30a**), **DMAC-BP** (**30b**), and **DTCz-BP** (**30c**) were determined as 716, 648, and 561 nm. The fluorescence spectroscopy data for **PXZ-BP** (**30a**) and **DMAC-BP** (**30b**) are only shown as excerpts. While conducting data processing, artifact bands were visible due to the low intensity of the photoluminescence of the compounds when the toluene blanc measurement was subtracted from the sample measurement. Therefore, only excerpts highlight the maximum emission wavelength, while the full fluorescence spectra are shown in Figure 59 (chapter 8.1) compared to the respective toluene blanc measurements. As expected, a stronger donor led to a red-shifted emission.

*

Lastly, the MOF-linker precursor **DTCz-BP-2R** (**35**) was investigated using UV-vis and fluorescence spectroscopy. The results thereof are displayed in Figure 14.

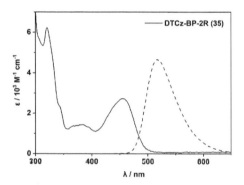

Figure 14. UV-vis and fluorescence (dotted lines) spectroscopy results for **DTCz-BP-2R** (**35**) measured in 0.1 mM toluene solution.

Absorption bands can be observed at 321, 382, and 456 nm with extinction coefficients of 6.22, 1.43, and $2.73 \cdot 10^3$ M^{-1} s^{-1}, respectively. Furthermore, the maximum emission wavelength was observed at 518 nm.

*

The measurement results of the UV-vis and fluorescence spectroscopy experiments for the D-A-type BP-based fluorophores are summarized in Table 1.

Table 1. Summary of the UV-vis and fluorescence spectroscopy data for the D-BP series (**30a-c**) and the MOF-linker precursor **2DTCz-BP-2R** (**35**) measured in 0.1 mM toluene solution.

Entry	Compound	λ_{max} (ε) [nm] ([10^3 M^{-1} cm^{-1}])	λ_{em} (λ_{exc}) [nm]
1	PXZ-BP	312 (50.11), 378 (28.27). 399 (36.21), 495 (0.31)	716 (495)
2	DMAC-BP	312 (14.15), 378 (9.17), 399 (11.75), 488 (0.17)	648 (490)
3	DTCz-BP	312 (53.91), 377 (21.78), 398 (27.99), 468 (2.76)	561 (467)
4	DTCz-BP-2R	321 (6.22), 382 (1.43), 456 (2.73)	518 (420)

As indicated in entries 1-3, the ICT absorption band of **PXZ-BP** (**30a**), **DMAC-BP** (**30b**), and **DTCz-BP** (**30c**) is bathochromically shifted in dependence on the respective donor strength, while the emission maxima range between 561-716 nm in toluene solution. Furthermore, the comparison between **DTCz-BP** (**30c**) (entry 3) and **DTCz-BP-2R** (**35**) shows that the shift of the DTCz donor unit from the 10-position to the 11-position in addition to the extended π-backbone, results in a blue-shift of the ICT absorption band, similar to the observations made by Yang and coworkers.[107] Furthermore, it is noteworthy that the extinction coefficients for the CT transition bands above 400 nm are in a narrow range. However, the values of ε differ quite starkly before 400 nm, in the spectral region, where π→π* transitions of the donor and acceptor, including locally excited (LE) vibronic absorption of the BP acceptor is usually observed.[106-107, 117, 120] Zhou et al. had reported BP-based D-A-type emitters that depicted vibronic absorption of the rigid BP moiety at 371 and 391 nm, which is in good accordance with the observations made for **PXZ-BP** (**30a**), **DMAC-BP** (**30b**) and **DTCz-BP** (**30c**) (entry 1-3).[118] **DTCz-BP-2R** (**35**) (entry 4), however, does not show these characteristic BP peaks, which indicates that the π-conjugated chain at the backbone of the fluorophore linker has a significant impact on the electronic structure of the molecule. Moreover, the emission wavelength is hypsochromically shifted for **DTCz-BP-2R** (**35**) compared to **DTCz-BP** (**30c**).

3.1.2. Dibenzo[*a,c*]phenazine-Based 2D-A-D- and 2D-A-R-Type Structures

2D-BP-D and 2D-BP-R-Type Structures - Synthesis

After investigating the D-A conjugates, (poly-donor)-acceptor structures were synthesized and characterized. Longwave thermally activated delayed fluorescence requires a highly stabilized ICT state. Thus, the desire to ensure a strong donor-acceptor interaction comes naturally.[121] The (poly-donor)-acceptor approach is a commonly applied strategy in investigations of TADF emitting materials and aims to increase the CT character.[36]

The substituted BP acceptor core was chosen to allow multiple donor substitutions. 3,6-Dibromo-10-fluorodibenzo[*a,c*]phenazine (**36**) was obtained in 92% yield as a beige solid by condensing dibrominated phenanthrene and diamine, as depicted in Scheme 8.

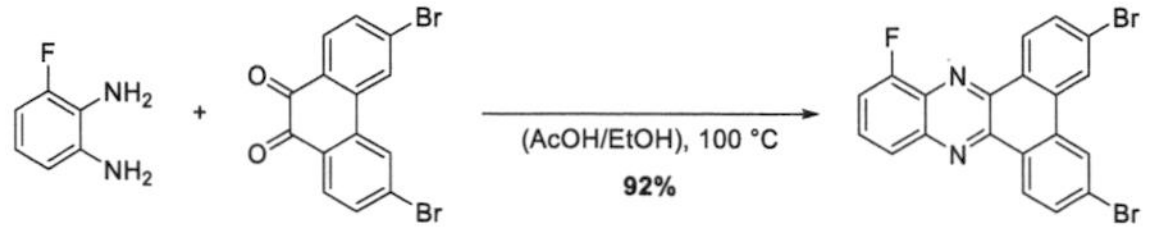

Scheme 8. Synthesis of precursor **x** for 2D-A-D-type structures.

Following the synthesis of **2Br-BP-F** (**36**), nucleophilic aromatic substitution (S$_N$Ar) with the 10*H*-phenoxazine **31a**, 9,9-dimethyl-9,10-dihydroacridine **31b**, and 3,6-di-*tert*-butyl carbazole **31c** donors was investigated as summarized in Table 2. **DTCz-BP-2Br** (**37c**) was obtained in 79% yield as a yellow solid. A structurally related TADF emitter to **DTCz-BP-2Br** (**37c**) exhibiting two nitrile substituents instead of the bromines, namely **tCz-BPCN**, was reported by Lee and coworkers in 2020 in a yield of 57%.[106] Attempts to synthesize **DMAC-BP-2Br** (**37b**) and **PXZ-BP-2Br** (**37a**) through nucleophilic aromatic substitution were unsuccessful under the reaction conditions given in Table 2.

Table 2. Nucleophilic aromatic substitution reactions leading to the synthesis of **D-BP-2Br** substrates (**37a-c**).

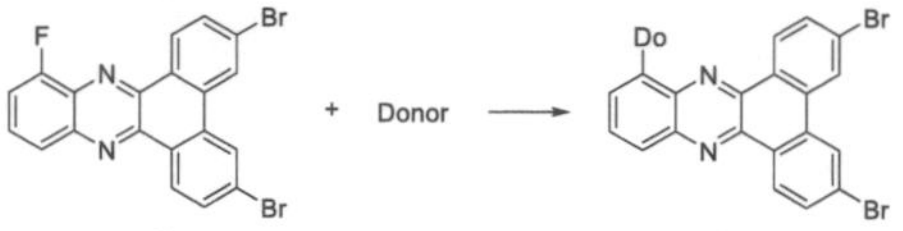

(36) 31a-c (37a-c)

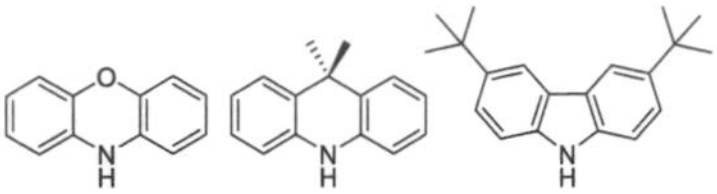

31a 31b 31c

Entry	Donor	Reaction conditions	Yield [%]
1	**31a**	K_3PO_4, (DMSO), 100 °C	-
2	**31b**	K_3PO_4, (DMSO), 100 °C	-
3	**31c**	K_3PO_4, (DMSO), 110 °C	79
4	**31a**	NaH, (DMF), 100 °C	-
5	**31b**	NaH, (DMF), 100 °C	-

The successfully synthesized intermediate (**37c**) was then used to synthesize **2PXZ-BP-DTCz** (**38**) in a Buchwald-Hartwig cross-coupling reaction with 10*H*-phenoxazine **31a** in a yield of 94% as a dark red solid, as displayed in Scheme 9.

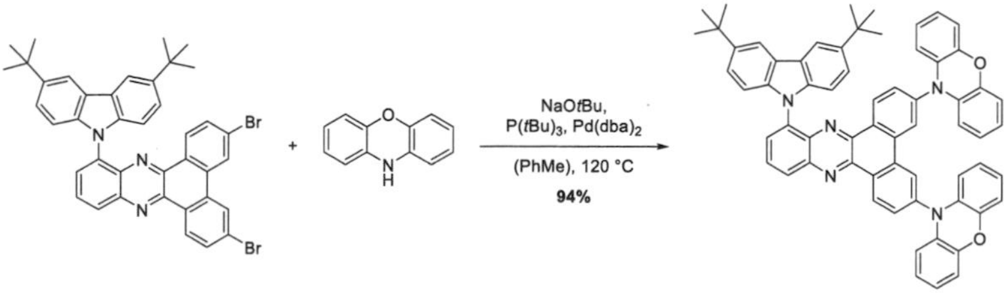

DTCz-BP-2Br (37c) **31a** **2PXZ-BP-DTCz (38)**

Scheme 9. Synthesis of 2D-A-D-type compound **2PXZ-BP-DTCz** (**38**).

As the syntheses of the intermediates **37a** and **37b** were unsuccessful by S_NAr, the synthesis strategy was adjusted to employ the Buchwald-Hartwig cross-coupling first, followed by S_NAr. Therefore, the target compounds after the cross-coupling starting from **2Br-BP-F** (**36**) were **2PXZ-BP-F** (**39a**), **2DMAC-BP-F** (**39b**) and **2DTCz-BP-F** (**39c**). These three molecules belonging to the 2D-BP-F series were successfully synthesized under the identical reaction conditions given in Table 3, entries 1-2, and 4.

Table 3. Reaction conditions and product yields of BUCHWALD-HARTWIG cross-couplings starting from **2Br-BP-F (36)**, Do = donor, PXZ = 10*H*-phenoxazine, DMAC = 9,9-dimethyl-9,10-dihydroacridine, DTCz = 3,6-di-*tert*-butyl carbazole.

Entry	Donor		[Pd]	Ligand	Δ [°C]	Product (Yield)
1	PXZ	**31a**	Pd(dba)$_2$	[(*t*Bu)$_3$PH]BF$_4$ (0.1 eq.)	120	**39a** (quant.)
2	DMAC	**31b**	Pd(OAc)$_2$	P(*t*Bu)$_3$ (0.2 eq.)	110	**39b** (90%), **40b** (10%)
3	DTCz	**31c**	Pd(OAc)$_2$	P(*t*Bu)$_3$ (0.2 eq.)	110	**39c** (-), **39d** (8%), **40c** (40%)
4	DTCz	**31c**	Pd(OAc)$_2$	P(*t*Bu)$_3$ (0.5 eq.)	120	**39c** (97%)

Table 3, entry 1 shows the synthesis of **2PXZ-BP-F (39a)** which was obtained as a red solid in a quantitative yield using a Pd(dba)$_2$/[(*t*Bu)$_3$PH][BF$_4$] catalyst system. In entry 2, the reaction conditions of the synthesis of **2DMAC-BP-F (39b)** are shown. The reaction was conducted with a Pd(OAc)$_2$/P(*t*Bu)$_3$ catalyst system and yielded the desired compound as an orange solid with a yield of 90%. The side product was also collected and characterized upon seeing a second band on the column. The side product **2DMAC-BP-O*t*Bu (40b)** was observed as an orange-yellow solid with a yield of 10% and showed an interesting phenomenon for the DMAC donors in the NMR analysis.

52

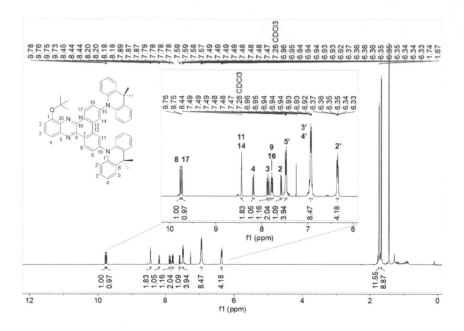

Figure 15. ¹H NMR of **2DMAC-BP-O*t*Bu (40b)** in CDCl₃.

Figure 15 shows the ¹H NMR of **2DMAC-BP-O*t*Bu (40b)** in CDCl₃. The proton signals could be assigned by consulting the 1D and 2D NMR data. While the NMR data assignment is fully enclosed in chapter 5.2.1, only the signals of the 9,9-dimethyl-9,10-dihydroacridinyl moieties will be discussed here. Due to the nitrogen atom's proximity and electron density, the 2'-H is expected to be the most shielded and show an upfield shift. Therefore, it was assigned to the multiplet between δ = 6.41−6.31 ppm. In the COSY NMR spectrum (Figure 16a), a cross-coupling peak of 2'-H with the multiplet between δ = 7.02−6.87 ppm is observed. Thus, 3'-H is assigned to this multiplet.

Subsequently, the assignment of 4'-H and 5'-H to the multiplet between δ = 7.02−6.87 ppm and δ = 7.51−7.45 ppm remains. The HMBC and HSQC NMR, as displayed in Figure 16b and Figure 16c, are analyzed. While 4'-H should be showing ³J coupling to C-2' and C-6' in the HMBC NMR spectrum, 5'-H is expected to exhibit ³J coupling to C-1', C-3', and Cq. The HMBC NMR shown in Figure 16b depicts a cross-peak between the multiplet between δ = 7.51−7.45 ppm and the ¹³C signal at δ = 36.17 ppm, which belongs to Cq. Therefore, 5'-H was assigned to this multiplet, and 4'-H to the shared multiplet with 3'-H.

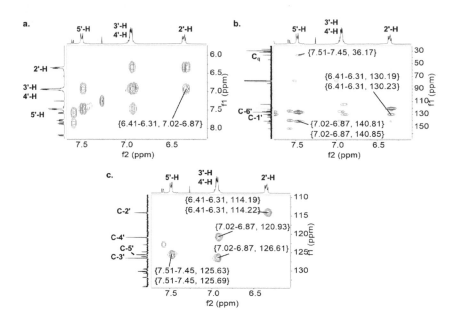

Figure 16. Excerpts from the 2D NMR experiments of **2DMAC-BP-O*t*Bu (40b)** in CDCl₃, **a.** COSY NMR, **b.** HMBC, and **c.** HSQC NMR.

Following the assignment of the DMAC protons, the assignment of their respective ¹³C signals was conducted. Through HSQC NMR (Figure 16), C-5' and C-2' were assigned to δ = 125.69, 125.63 ppm and δ = 114.22, 114.19 ppm, respectively. To identify C-3' and C-4', HMBC NMR was used to find the cross-peaks with 5'-H and 2'-H, respectively. Although not marked in Figure 16b due to overview reasons, C-3' and C-4' signals were found at δ = 126.61 and 120.93 ppm, respectively. Then, the signals for C-1' and C-6' were left to be assigned. C-1' is expected to be downfield shifted in comparison to C-6' due to the electronegativity difference of N > C. Thus, based on this and with the aid of HMBC NMR that showed the cross-peaks to 3'-H and 4'-H, respectively, C-1' was identified as δ = 140.85, 140.81 ppm and C-6' as δ = 130.23, 130.19 ppm. Lastly, the methyl groups were assigned employing HSQC NMR to δ = 31.85, 31.75 ppm.

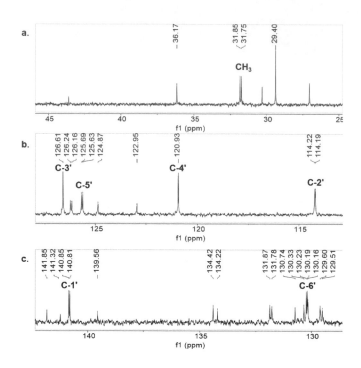

Figure 17. Excerpts of the ^{13}C NMR of **2DMAC-BP-O*t*Bu** (**40b**) in CDCl$_3$, panel **a.** shows impurities at δ = 43.61 ppm, 30.32 ppm and δ = 27.05 ppm stem from cyclohexane and an unidentified impurity.

Interestingly, not all DMAC protons are doubled, but solely C-1', C-2', C-5', C-6' and the methyl groups. The MS analysis confirmed the product formation and unreacted DMAC was ruled out based on literature-reported NMR data.[122] Therefore, the observation of rotamers was considered. However, it remains unclear if the C$_q$, C-3', and C-4' signals are identical for the assumed rotamers and, if that is the case, why they are unaffected by the conformational isomerism.

When the same reaction conditions as shown in Table 3, entry 1 were applied to the cross-coupling of DTCz to 2Br-BP-F, the desired product (**39c**) could not be observed. Instead, the tri-donor-substituted **2DTCz-BP-DTCz** (**39d**) and the *tert*-butoxy-substituted **2DTCz-BP-O*t*Bu** (**40c**) were obtained as an orange solid and yellow film with yields of 8 and 40%, respectively. Thereby, the second 2D-A-D-type compound (**39d**) was obtained. Increasing the ligand amount from 0.20 equiv. to 0.50 equiv. then successfully led to the synthesis of **2DTCz-BP-F** (**39c**) as a yellow crystalline solid with a yield of 97%.

The formation of **2DMAC-BP-O*t*Bu** (**40b**) and **2DTCz-BP-O*t*Bu** (**40c**) is literature-known. Palladium catalyst systems catalyzing C-O bond formation starting from aryl halides had been extensively reported by Hartwig and coworkers, as well as Buchwald and coworkers in the late 90s.[123-127] Building on their findings, Watanabe *et al.* continued to investigate the Pd-catalyzed synthesis of *tert*-butyl aryl ethers using P(*t*Bu)$_3$ as a ligand.[128] They reported that tri-*tert*-butyl phosphine showed high activity in yielding aryl ethers from aryl bromides and aryl chlorides, but the synthesis of aryl ethers from aryl fluorines with tri-*tert*-butyl phosphine as a ligand was not mentioned. This is reasonable as activating the very strong C-F bond through oxidative addition is not as effective. Hartwig and coworkers continued to report similar findings using a Pd(dba)$_2$/Ph$_5$FcP(*t*Bu)$_2$ catalyst system in 2000[129] and 2002[130].

Based on these syntheses results, it was attempted to find a synthesis pathway to **2PXZ-BP-O*t*Bu** (**40a**), as this substrate was not observed as a side product for the formation of **2PXZ-BP-F** (**39a**) (Table 3, entry 1). Although the synthesis starting from the respective aryl bromides or aryl chlorides would have been more obvious choices, this was not compatible with the current modular synthesis strategy using a 2Br-BP-X-type starting material because unselective substitution would have occurred when installing either the donors or the *tert*-butoxy substituent. Therefore, an aryl fluorine was selected as the starting material. Furthermore, this aligns with the assumed synthesis pathway for its DMAC- and DTCz-featuring analogs. The compound **2PXZ-BP-O*t*Bu** (**40a**) was synthesized from compound (**39a**) under the reaction conditions shown in Scheme 10 as an orange solid with a yield of 34%. Even with a reaction duration of six days, the starting material was recovered in approximately 38% yield.

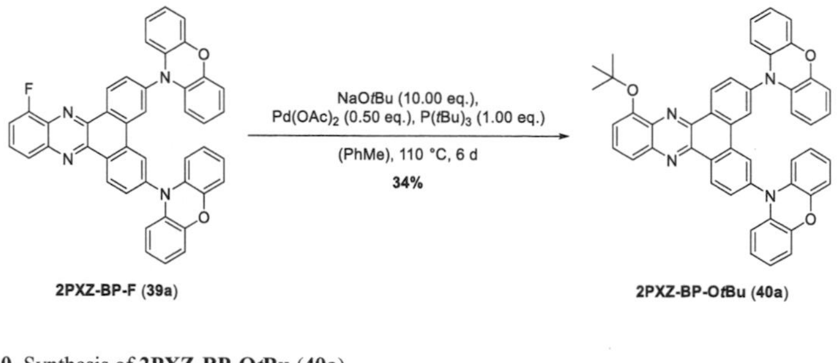

NaO*t*Bu (10.00 eq.),
Pd(OAc)$_2$ (0.50 eq.), P(*t*Bu)$_3$ (1.00 eq.)

(PhMe), 110 °C, 6 d

34%

2PXZ-BP-F (39a)

2PXZ-BP-O*t*Bu (40a)

Scheme 10. Synthesis of **2PXZ-BP-O*t*Bu** (**40a**).

The synthesis route first attaching the two donor units to the acceptor core before substituting at the fluorine position was successfully established. Using this synthesis route, three 2D-BP-F emitters were successfully synthesized. Moreover, it was shown that fluorine substitution with a *tert*-butoxy substituent occurred under BUCHWALD-HARTWIG cross-coupling conditions using Pd(OAc)$_2$/P(tBu)$_3$ as the catalyst system and NaOtBu as a base. Three compounds were successfully obtained as examples for the 2D-BP-OtBu series for this observation. However, this unforeseen side reaction showed shortcomings regarding the yields and very long reaction times. On the other hand, it also revealed the opportunity to re-think and improve the synthesis strategy of BP-based TADF emitters. Consequently, the following synthesis strategy displayed in Figure 18 was investigated.

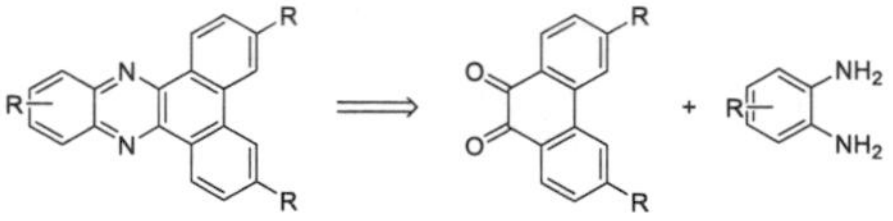

Figure 18. Retrosynthetic scheme to an additional synthesis strategy towards BP-based TADF emitters **X** employing phenanthren-9,10-dione derivatives **28c-e** and 1,2-diaminobenzene derivatives **27c-f**.

The series of TADF molecules based on the dibenzo[*a,c*]phenazine acceptor was thus continued through a different synthesis strategy employing a condensation reaction between phenanthren-9,10-dione derivatives **28c-e** and 1,2-diaminobenzene derivatives **27c-f** as the final step. The charm of this approach lies in the commercial availability of 1,2-diaminobenzenes **27c-f** through which a facile build-up of a 2D-BP-R series can be achieved that focuses on the derivatization of the BP acceptor core.

To synthesize the phenanthrene-9,10-dione derivatives, the carbonyl groups were masked to facilitate the handling. Thus, the 3,6-dibromophenanthrene-9,10-dione **28b** was first protected with a cyclic acetal group to give the protected compound **41**. Then, the donor moieties DTCz, DMAC, and PXZ were attached *via* BUCHWALD-HARTWIG amination. Subsequently, compounds **42a-c** were deprotected under acidic conditions at 120 °C to give **28c-e** in moderate to excellent yields, as shown in Scheme 11. The overall yield over three steps for **28c**, **28d**, and **28e** amounted to 51, 61, and 37%, respectively. While compound **28c** was reported by Adachi and coworkers[108] and Zhang and coworkers[105, 131], compound **28d** was filed in a patent by

Ma *et al*[132] in 2021. On the other hand, compound **28e** was not directly reported in the literature, but two patents mention the unsubstituted carbazole derivatives.[119, 132]

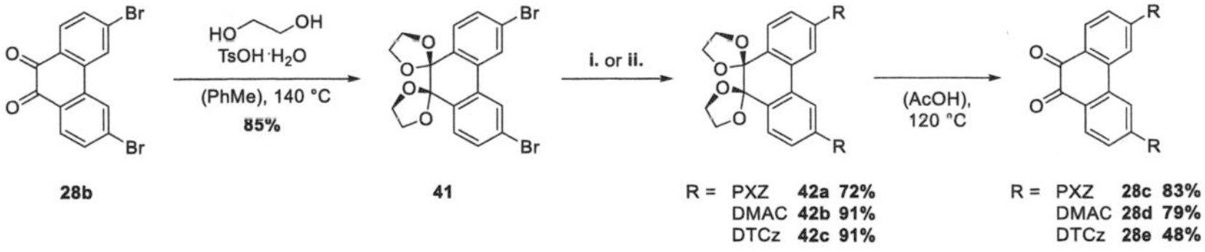

Scheme 11. Synthesis pathway of donor-coupled phenantrene-9,10-dione derivatives **28c-e**; **i.** PXZ, NaO*t*Bu, Pd(dba)₂, [(*t*Bu)₃PH]BF₄, (PhMe), 120 °C, **ii.** DMAC or DTCz, NaO*t*Bu, Pd(OAc)₂, P(*t*Bu)₃, (PhMe), 120 °C.

After the building blocks **28c-e** were successfully obtained, commercially available 1,2-diaminobenzenes with various substituents were used to yield compounds **43c-e** and **44c-f** in condensation reactions in *n*-butanol under reflux at 120 °C as shown in Scheme 12 and Scheme 13.

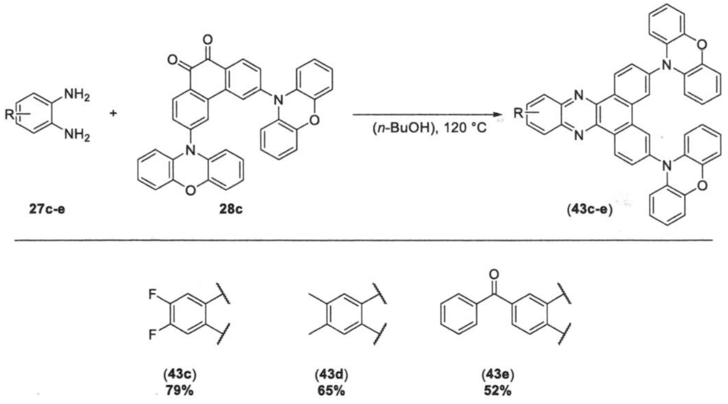

Scheme 12. Dibenzo[*a,c*]phenazine derivatives (**43c-f**) featuring PXZ donors.

Following the design rationale of the 2D-BP-F series, an additional fluorine substituent was introduced to stabilize the acceptor further.[44] The methyl groups were anticipated not to induce a drastic change in the LUMO but slightly destabilize the LUMO based on their weak electron-donating nature. Furthermore, the idea of the 11-benzoyl-dibenzo[*a,c*]phenazine acceptor was a product of fusing two known TADF acceptors, namely dibenzo[*a,c*]phenazine and benzophenone. In 2016, Dey and coworkers had reported the 11-benzoyl-dibenzo[*a,c*]phenazine system as an environment sensitive fluorescent probe that acts *via* excited state hydrogen bonding.[133] The addition of donors was expected to lead to efficient intramolecular charge transfer and a structural twist and TADF properties. Compounds (**43c-f**) were obtained as red to dark-red colored solids in overall yields over four steps of 30, 40, and 27%, respectively. **2PXZ-BP-2F** (**43c**) has been reported in several patents[134-137], while **2PXZ-BP-2Me** (**43d**) has not been reported thus far. Furthermore, **2PXZ-BP-COPh** (**43e**) was published by Wang and coworkers in 2020 and showed excellent TADF properties and high device performance, affirming the molecule design displayed in Scheme 12.[138]

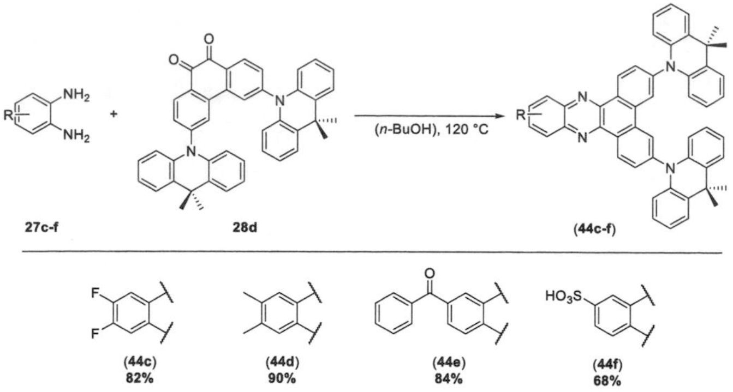

Scheme 13. Dibenzo[*a,c*]phenazine derivatives (**44c-f**) featuring DMAC donors.

Compounds **44c-f** were obtained as yellow, orange to red-colored solids in overall yields over four steps of 50, 55, 51, and 41%, respectively. As of 04/20/2022, a Scifinder[n] search did not indicate that **2DMAC-BP-2F** (**44c**), **2DMAC-BP-2Me** (**44d**), and **2DMAC-BP-SO3H** (**44f**) had been published so far. However, **2DMAC-BP-COPh** (**44e**) was recently published by Zhao and coworkers in 2022,[139] and they successfully filed a patent.[140] In their paper, they report excellent TADF properties and very good OLED device performance.

Although some of these compounds were published, the synthesis strategy leading to facile derivatization of the BP acceptor core was successfully established. Furthermore, **2DMAC-BP-SO3H** (**44f**) is especially interesting regarding its potential application in cellular imaging as the sulfonic acid functional group increases the hydrophilicity. The BP acceptor core has received increased attention over the past years and its applications are now expanded to aqueous systems. For example, Jiang and coworkers recently reported a distantly related structure to **2DMAC-BP-SO₃H** (**44f**) featuring a 2,4-dinitrobenzenesulfonyl group as thiophenol sensors in environmental water.[141]

Three modular synthesis pathways leading to 2D-BP-D-type and 2D-BP-D-R-type compounds were successfully established. While the first one employed an S_NAr reaction with **2Br-BP-F** (**36**) followed by BUCHWALD-HARTWIG coupling, the second pathway investigated the direct BUCHWALD-HARTWIG coupling to the **2Br-BP-F** (**36**) substrate. The first pathway yielded the 2D-BP-D-type compound **2PXZ-BP-DTCz** (**38**), whereas the second synthesis strategy led to the 2D-BP-F series comprising **2PXZ-BP-F** (**39a**), **2DMAC-BP-F** (**39b**) and **2DTCz-BP-F** (**39c**).

One side reaction observed for the second synthesis pathway was the trisubstitution leading to the formation of **2DTCz-BP-DTCz** (**39d**), whereas another side reaction led to the replacement of the fluorine substituent with a *tert*-butoxide substituent, while cross-coupling donor moieties to the acceptor core. Thereby, the 2D-BP-O*t*Bu series containing **2PXZ-BP-O*t*Bu** (**40a**), **2DMAC-BP-O*t*Bu** (**40b**), and **2DTCz-BP-O*t*Bu** (**40c**) was obtained. This unwanted side reaction was effectively prevented through the third synthesis strategy, whose key feature was the use of a condensation reaction in the final step. Moreover, the third pathway enabled easy access to the derivatization of the BP acceptor unit. Substrates with two fluorine substituents (e.g., 2D-BP-2F), two methyl substituents (e.g., 2D-BP-2Me), or a benzoyl unit (e.g., 2D-BP-COPh) attached to the BP core were obtained without much synthetic effort.

2D-BP-D and 2D-BP-R-Type Structures – DFT Calculations and Electrochemical Behavior

After the synthesis of the (poly-donor)-acceptor conjugates were conducted, TD-DFT calculations and cyclic voltammetry were carried out to investigate the donor and acceptor properties and assess the TADF potential of the compounds. The computational experiments were kindly provided by Changfeng Si from the Zysman-Colman group at the University of St. Andrews and were carried out in the gas phase using the Pbe1pbe/6-31G(d,p) functional. The ground-state geometries of the respective molecules were optimized using DFT and the excited states were computed using the time-dependent DFT within the TAMM-DANCOFF approximation (TDA) based on the optimized ground-state geometries.

In this subchapter, the TD-DFT calculation results in combination with the obtained CV data for the 2D-BP-D, 2D-BP-F, 2D-BP-OtBu and 2D-BP-R series will be presented separately in the depicted order. Subsequent to each data explanation, a comparative discussion of each series results follows. At the end of the subchapter, an overall comparison of the TD-DFT results and CV results of the four molecule series is considered.

*

The initial motivation for the synthesis of the 2D-BP-D substrates had been the realization of the (poly-donor)-acceptor strategy which aims for an increased CT character. The results of the TD-DFT calculation of the 2D-BP-D series are shown in Figure 19 and summarized in Table 4, entries 1-2. The geometry-optimized ground state of **2PXZ-BP-DTCz (38)** shows that the HOMO is located on one of the PXZ donors, while the LUMO is found on the BP acceptor. For **2DTCz-BP-DTCz (39d)**, on the other hand, it is observed that the HOMO is located mainly on the DTCz donor, which was introduced by $S_{N}Ar$, and to a lesser extent on a DTCz donor introduced by BUCHWALD-HARTWIG coupling, while the LUMO is also found on the BP acceptor. Interestingly, in the case of **2PXZ-BP-DTCz (38)**, the third DTCz donor, introduced onto the BP core *via* $S_{N}Ar$, does not actively contribute to neither the HOMO nor the LUMO. A similar observation was made by Zhao and coworkers, who reported a tri-PXZ-substituted TADF emitter called **3PXZ-BP (3c)** for which the DFT calculations displayed the same phenomenon. They mention that the third PXZ donor plays a role in the HOMO-2→LUMO transition requiring more energy.[42]

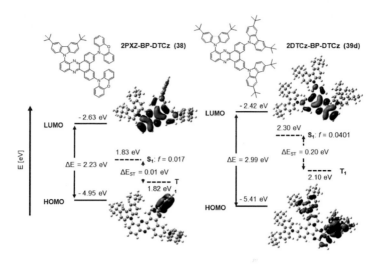

Figure 19. Geometry optimization and excited state TD-DFT calculation results of the 2D-BP-D series provided by Changfeng Si (Zysman-Colman Group, University of St. Andrews); computation in the gas phase using the Pbe1pbe/6-31G(d,p) functional.

Furthermore, PXZ (E_{HOMO} = -4.95 eV) shows a shallower HOMO level than DTCz (E_{HOMO} = -5.41 eV) following the reported donor strength.[36] Moreover, the LUMO level of **2PXZ-BP-DTCz (38)** (E_{LUMO} = -2.63 eV) is stabilized in comparison to **2DTCz-BP-DTCz (39d)** (E_{LUMO} = -2.42 eV). This results in a HOMO-LUMO energy gap of 2.23 eV and 2.99 eV for **2PXZ-BP-DTCz (38)** and **2DTCz-BP-DTCz (39d)**, respectively. The excited-state calculation results show that the nature of the S_1 state of **2PXZ-BP-DTCz (38)** is represented by a HOMO→LUMO (99%) transition, while the T_1 state thereof is also a HOMO→LUMO (98%) transition. The calculation for **2DTCz-BP-DTCz (39d)** shows that the S_1 state consists of an 88% HOMO→LUMO transition, while the 82% of the nature of the T_1 state stems from the HOMO→LUMO transition. Furthermore, the excited states S_1 and T_1 of **2PXZ-BP-DTCz (38)** are energetically lower at 1.83 eV and 1.82 eV, respectively, in comparison to **2DTCz-BP-DTCz (39d)** for which the S_1 and T_1 were calculated to 2.30 eV and 2.10 eV. This data results in a ΔE_{ST} of 0.01 eV and 0.20 eV for **2PXZ-BP-DTCz (38)** and **2DTCz-BP-DTCz (39d)**, respectively. Both lie in a range where RISC could occur efficiently at room temperature, although **2PXZ-BP-DTCz (38)** is anticipated to show superior TADF properties due to its minuscule ΔE_{ST}. However, as the S_1 state from which fluorescence is emitted is comparatively low-lying, it is expected that **2PXZ-BP-DTCz (38)** shows red-shifted emission compared to **2DTCz-BP-DTCz (39d)**. As an additional consequence of the low-lying S_1 state, the probability of competing for non-radiative de-excitation processes increases.

Subsequent to the DFT calculations, cyclic voltammetry was conducted to obtain experimental values for the HOMO and LUMO energy levels. The measurements for the 2D-BP-D series were carried out in DMF due to severe solubility issues of **2DTCz-BP-DTCz (39d)** in DCM. However, **2DTCz-BP-DTCz (39d)** could also not be fully dissolved in DMF either and therefore the solution was filtered over cotton before it was submitted to potentiodynamic analysis. Information on the measurement setup and data processing can be found in chapter 5.1. The CV measurements for the 2D-BP-D series are displayed in Figure 20 and summarized in Table 5, entries 1-2.

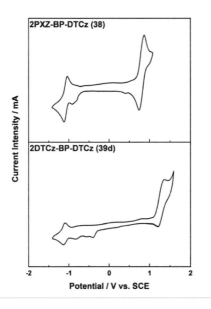

Figure 20. Cyclic voltammogram of the 2D-BP-D-type **2DTCz-BP-DTCz (39d)** and **2PXZ-BP-DTCz (38)** measured in Ar-saturated DMF (0.1 M [nBu$_4$N][PF$_6$]) at a scan rate of 100 mV s^{-1}.

Both compounds show multiple reduction processes and additional signals at positive potentials. While the oxidation process of **2PXZ-BP-DTCz (38)** is fully reversible, the oxidation process in **2DTCz-BP-DTCz (39d)** shows quasi-reversible behavior. The oxidation potentials for **2PXZ-BP-DTCz (38)** and **2DTCz-BP-DTCz (39d)** were 0.80 and 1.29 V vs. SCE, respectively.

The main reduction processes of **2PXZ-BP-DTCz (38)** and **2DTCz-BP-DTCz (39d)** were reversible, and the reduction potentials amounted to -1.08 and -1.11 V vs. SCE, respectively. However, at negative potentials in the voltammogram of **2PXZ-BP-DTCz (38)**, other signals in the cathodic trace were observed at E_{red}^c = -0.81 and -0.91 V vs. SCE, with hinted opposite peaks in the anodic trace at E_{red}^a = -0.61 and -0.83 V vs. SCE, respectively.

Based on the main oxidation and reduction signals, the HOMO and LUMO energies were calculated to be -5.14 eV and -3.26 eV for **2PXZ-BP-DTCz (38)**, respectively, and -5.63 eV and -3.23 eV for **2DTCz-BP-DTCz (39d)**, respectively. The HOMO-LUMO gap thus amounts to 1.89 and 2.40 eV for **2PXZ-BP-DTCz (38)** and **2DTCz-BP-DTCz (39d)**, respectively.

A literature search was carried out to find structurally related compounds for comparison. Unfortunately, the literature reported HOMO and LUMO values for **2PXZ-BP (3b)** and **3PXZ-BP (3c)**, which are structurally related to **2PXZ-BP-DTCz (38)**, could not be used for comparison, as Zhao and co-workers had measured the CV in DCM at lower scan rate and Chen *et al.* had not provided information for their CV measurement of **DBPZ-DPXZ** regarding the scan rate or measurement setup.[42, 105] A structurally related carbazole-bearing BP derivative was reported by Singh *et al.* in 2021, shown in Figure 21. They determined the HOMO and LUMO as -5.56 and -3.03 eV, respectively.[142] In comparison to **2DTCz-BP-DTCz (39d)**, the HOMO of **Dye 4 (45)** is slightly lower, while the LUMO is more stabilized.

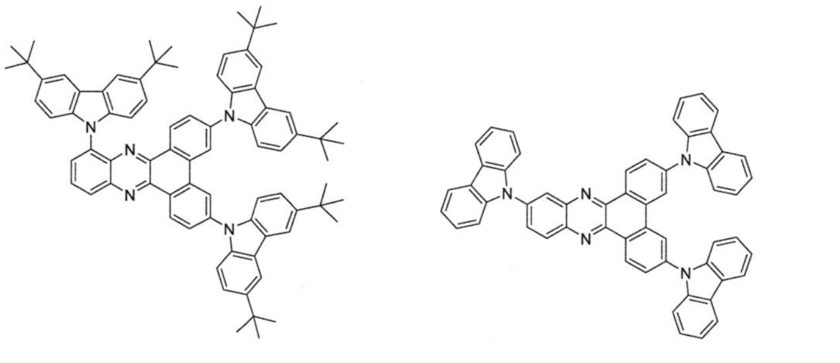

2DTCz-BP-DTCz (39d)

Dye 4 (45)
Singh *et al.*, *J. Photochem. Photobiol. A: Chem.* **2021**, *419*, 113457.

Figure 21. Molecule structures of **2DTCz-BP-DTCz (39d)** and **Dye 4 (45)** reported by Singh *et al.*[142]

In comparison to each other, **2PXZ-BP-DTCz (38)** displays a shallower HOMO level than **2DTCz-BP-DTCz (39d)**, following the respective donor strength.[36] Furthermore, the LUMO energies lie in a close range. However, both compounds show many other processes that imply that electrons are not only transferred in one defined transaction from a donor to an acceptor but also multiple processes occur. Also, the DFT calculations for the **2PXZ-BP-DTCz (38)** showed that the third donor (DTCz) does not actively contribute to the HOMO or LUMO. This finding indicates that a leaner molecule design could effectively yield TADF emitters demonstrating strong ICT.

*

The reduced molecule design of the 2D-BP-F series comprising **2PXZ-BP-F (39a)**, **2DMAC-BP-F (39b)**, and **2DTCz-BP-F (39c)** was successfully established and published.[40] The results of the TD-DFT calculations are shown in Figure 22 and summarized in Table 4, entries 3-5.

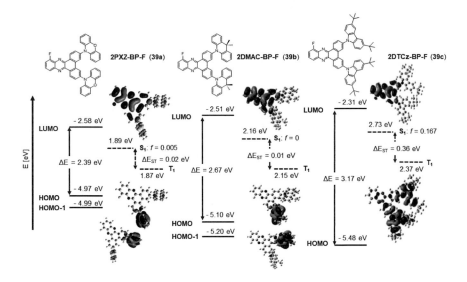

Figure 22. Geometry optimization and excited state TD-DFT calculation results of the 2D-BP-F series (**39a-c**) provided by Changfeng Si (Zysman-Colman Group, University of St. Andrews); computation in the gas phase using the Pbe1pbe/6-31G(d,p) functional.

The HOMO levels are located on the donors in all three emitters, while the LUMO representing the acceptor is found on the BP moiety. Under the expected donor strength, **2PXZ-BP-F (39a)** shows the shallowest HOMO, while **2DTCz-BP-F (39c)** shows the lowest HOMO level in this triade. Furthermore, the LUMO levels lie in a similar range, although it is less stabilized for **2DTCz-BP-F (39c)** than in **2PXZ-BP-F (39a)**. The HOMO-LUMO gaps amount to 2.39, 2.67, and 3.17 eV for **2PXZ-BP-F (39a)**, **2DMAC-BP-F (39b)**, and **2DTCz-BP-F (39c)**, respectively. The dihedral angle defining the structural twist is a crucial design parameter to minimize ΔE_{ST}, as mentioned in previous chapters. For **2PXZ-BP-F (39a)**, **2DMAC-BP-F (39b)**, and **2DTCz-BP-F (39c)**, the calculated dihedral angles amount to 85.9° and 72.2°, 87.9° and 90.1°, and 48.6° and 47.6°, respectively. Crystal structures for **2DMAC-BP-F (39b)** and **2DTCz-BP-F (39c)** were obtained (chapter 5.3) and showed a smaller dihedral angle of 65° for the **2DMAC-BP-F (39b)** and a dihedral angle of 46.6° for **2DTCz-BP-F (39c)** which was in good accordance with the calculated value. The large dihedral angles do not only result in spatial separation of the frontier molecular orbital, but also lead to a small ΔE_{ST} of 0.02 eV and 0.01 eV for **2PXZ-BP-F (39a)** and **2DMAC-BP-F (39b)**, respectively, but also a larger value of 0.36 eV for **2DTCz-BP-F (39c)**, which had displayed a smaller dihedral angle. The oscillator strengths were calculated as 0.005, 0, and 0.167 for **2PXZ-BP-F (39a)**, **2DMAC-BP-F (39b)**, and **2DTCz-BP-F (39c)**, respectively.

Subsequently, the electrochemical behavior of the 2D-BP-F series was investigated by means of cyclic voltammetry. The cyclic voltammograms are shown in Figure 23, and the results are summarized in Table 5, entries 3-5.

As shown in Figure 23, all three compounds show reversible oxidation and reduction processes. The main oxidation waves for **2PXZ-BP-F (39a)**, **2DMAC-BP-F (39b)**, and **2DTCz-BP-F (39c)** were observed at 0.80, 1.00, and 1.32 V vs. SCE, respectively. These were each assigned to the oxidation of PXZ, DMAC, and DTCz, and exhibited the relative strength of the donor units. **2DMAC-BP-F (39b)** shows an additional minor oxidation wave at 0.77 V vs. SCE, which is characteristic for the redox behavior of DMAC-containing compounds.[106] The HOMO levels were found at -5.14, -5.34, and -5.66 eV for **2PXZ-BP-F (39a)**, **2DMAC-BP-F (39b)**, and **2DTCz-BP-F (39c)**, respectively. The reductions of these three emitters occurred at very similar potentials of 1.21, 1.19, and 1.18 V vs. SCE, respectively, and match well with the reported reduction of the dibenzo[a,c]phenazine-based acceptor (-1.19 V vs. SCE)[107] and indicate that the electronic coupling between the donor and acceptor unit is small.

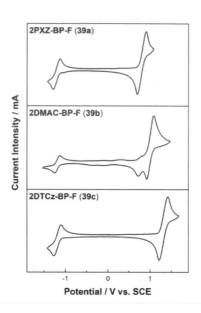

Figure 23. Cyclic voltammograms of **2PXZ-BP-F** (**39a**), **2DMAC-BP-F** (**39b**), and **2DTCz-BP-F** (**39c**) in Ar-saturated DCM solution (0.1 M [*n*Bu₄N][PF₆]) at a scan rate of 100 mV s⁻¹.

The fluorine substituent, which is attached to the BP acceptor, stabilizes the LUMO for **2PXZ-BP-F** (**39a**) (-3.13 eV), **2DMAC-BP-F** (**39b**) (-3.15 eV), and **2DTCz-BP-F** (**39c**) (-3.16 eV) in comparison to their non-fluorinated analogues **2PXZ-BP** (**2c**) (-2.63 eV)[42], **2DMAC-BP** (**2b**) (-2.49 eV)[43] and 2,7-bis(3,6-di-*tert*-butyl-9*H*-carbazol-9-yl)dibenzo[*a*,*c*]phenazine (**CzDbp**) (-2.5 eV)[113] by 0.50 eV, 0.65 eV and 0.66 eV, respectively. Thus, the presence of fluorine increases the acceptor strength. Lastly, the corresponding redox gaps, $\Delta E_{\text{HOMO-LUMO}}$, decrease from 2.50 to 2.20 and 2.01 V for **2DTCz-BP-F** (**39c**), **2DMAC-BP-F** (**39b**), and **2PXZ-BP-F** (**39a**), respectively. This is in accordance with the HOMO-LUMO gap trend predicted by the DFT calculations.

*

Next, the TD-DFT computation was carried out for the 2D-BP-OtBu series by Changfeng Si (Zysman-Colman Group, University of St. Andrews). The results are shown in Figure 24 and Table 4, entries 6-8. The ground-state geometry optimization shows that the LUMOs of **2PXZ-BP-O*t*Bu** (**40a**), **2DMAC-BP-O*t*Bu** (**40b**), and **2DTCz-BP-O*t*Bu** (**40c**) are mainly located on the BP acceptor core and slightly spread out onto the oxygen atom of the *tert*-butoxy substituent. On the other hand, the HOMOs are found on both donor units, although to different degrees. While the HOMO of **2DMAC-BP-O*t*Bu (40b)** seems to be equally distributed onto both DMAC donors, this is not the case for **2PXZ-BP-O*t*Bu (40a)** and **2DTCz-BP-O*t*Bu (40c)** with their respective donor moieties. This is also the case for the HOMO-1 of **2PXZ-BP-O*t*Bu (40a)** and **2DTCz-BP-O*t*Bu (40c)**. The HOMO-2 of **2DMAC-BP-O*t*Bu (40b)** is located on the BP core and is extensively extended onto the *tert*-butoxide moiety. This is reasonable as alkoxy substituents are known to belong to the activating groups that tend to donate electron density. The same phenomenon can be observed for the HOMO-4 of **2DTCz-BP-O*t*Bu (40c)**.

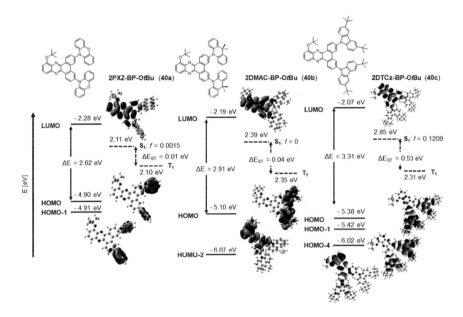

Figure 24. Geometry optimization and excited state TD-DFT calculation results for the 2D-BP-O*t*Bu series (**40a-c**) provided by Changfeng Si (Zysman-Colman Group, University of St. Andrews); computation in the gas phase using the Pbe1pbe/6-31G(d,p) functional.

The S_1 states of **2PXZ-BP-O*t*Bu (40a)**, **2DMAC-BP-O*t*Bu (40b)** and **2DTCz-BP-O*t*Bu (40c)** are calculated to an energy level of 2.11, 2.39 and 2.85 eV, respectively. According to Kasha's rule, radiative decay is only observed from the lowest-lying state. Therefore, the S_1 state is somewhat indicative of the emission properties. Following the calculated energy levels, **2PXZ-BP-O*t*Bu (40a)** is thus expected to emit at longer wavelengths than **2DTCz-BP-O*t*Bu (40c)**. Furthermore, the TD-DFT calculation showed that the nature of the S_1 state of

2PXZ-BP-O*t*Bu (**40a**) and **2DTCz-BP-O*t*Bu** (**40c**) is a mixture of HOMO-1→LUMO (37%) and HOMO →LUMO (61%), and a mixture of HOMO-1→LUMO (41%) and HOMO→LUMO (57%), respectively. In contrast, the nature of the S_1 state of **2DMAC-BP-O*t*Bu** (**40b**) was computed to be a HOMO→LUMO (90%) transition. The TD-DFT calculations also delivered data on the T_1 state of **2PXZ-BP-O*t*Bu** (**40a**), **2DMAC-BP-O*t*Bu** (**40b**) and **2DTCz-BP-O*t*Bu** (**40c**). The energies were calculated to be 2.10, 2.35, and 2. 31 eV, respectively. Furthermore, the nature of the T_1 state of **2PXZ-BP-O*t*Bu** (**40a**) was computed as a HOMO→LUMO (96%) transition, while for **2DMAC-BP-O*t*Bu** (**40b**), it was a HOMO-2→LUMO (91%) transition, and for **2DTCz-BP-O*t*Bu** (**40c**) it showed a mixed character involving a HOMO-4→LUMO (58%), HOMO-1→LUMO (21%) and HOMO→LUMO (11%) transition.

Following the data obtained from the TD-DFT calculations of the excited states, the ΔE_{ST} of **2PXZ-BP-O*t*Bu** (**40a**), **2DMAC-BP-O*t*Bu** (**40b**), and **2DTCz-BP-O*t*Bu** (**40c**) amounts to 0.01, 0.04 and 0.53 eV, respectively. From this it follows that the first two emitters have a good probability of showing efficient RISC and thereby TADF, while the last emitter is prone to show inefficient RISC.

As the **2DMAC-BP-O*t*Bu** (**40b**) and **2DTCz-BP-O*t*Bu** (**40c**) were obtained in low amounts, the electrochemical investigation of the 2D-BP-O*t*Bu series was represented by a CV measurement of **2PXZ-BP-O*t*Bu** (**40a**), for which sufficient amounts of materials were obtained. The result is shown in Figure 25 and Table 5, entry 6. Both the oxidation and reduction processes are reversible. The HOMO energy was calculated based on the oxidation potential to amount to -5.17 eV, while the LUMO energy was obtained as -3.09 eV. Thus, the HOMO-LUMO energy gap amounts to 2.07 eV. Additional reduction processes are observed in the cathodic trace at E_{red}^c = -1.20 V vs. SCE and in the anodic trace at E_{red}^a = -1.29 V vs. SCE. As these signals were not observed with **2PXZ-BP-F** (**39a**), it is suspected that these are reduction processes related to the *tert*-butoxide group.

*

Lastly, the 2PXZ-BP-R and 2DMAC-BP-R series obtained from the synthesis route aimed at facile acceptor modification were measured by means of CV to investigate the substituent effect on the LUMO of the acceptor core.

The oxidation and reduction processes for the 2PXZ-BP-R series are reversible, as displayed in Figure 25. The experimental data show that the PXZ donor strength in **2PXZ-BP-2F** (**43c**), **2PXZ-BP-2Me** (**43d**), and **2PXZ-P-COPh** (**43e**) remains comparatively consistent as the respective oxidation potentials vary between 0.85, 0.80 and 0.87 V vs. SCE, and the HOMO levels between -5.19, -5.14 and -5.21 eV, respectively.

In contrast, the reduction potentials and the LUMO energy levels, which indicate the acceptor strength, display more variance (Table 5). The reduction potentials of **2PXZ-BP-2F** (**43c**), **2PXZ-BP-2Me** (**43d**) and **2PXZ-P-COPh** (**43e**) amount to -1.15, -1.36, and -0.99 V vs. SCE, respectively. Thus, the LUMO levels of **2PXZ-BP-2F** (**43c**), **2PXZ-BP-2Me** (**43d**) and **2PXZ-BP-COPh** (**43e**) were calculated to amount to -3.19, -2.98, and -3.35 eV, respectively. Therefore, as intended by the molecule design rational of this 2D-BP-R series, the LUMO level is considerably stabilized by adding the benzoyl group, while an analogous, but less pronounced effect is observed for the difluoro-substituted compound. The dimethyl-substituted compound shows the least stabilized LUMO energy level in this triade.

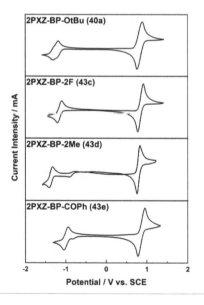

Figure 25. Cyclic voltammograms of the 2PXZ-BP-R series measured in Ar-saturated DCM containing 0.1 M [nBu_4N][PF_6] at a scan rate of 100 mV s^{-1}.

While **2PXZ-BP-2F** (**43c**) showed no additional reduction signals, **2PXZ-BP-2Me** (**43d**) and **2PXZ-BP-COPh** (**43e**) showed further peaks. For **2PXZ-BP-2Me** (**43d**), a reversible reduction process at -0.85 V vs. SCE was observed. Furthermore, the CV curve for **2PXZ-BP-2Me** (**43d**) shows a crossing of the anodic and cathodic trace, which is assumed to be due to a coating of the working electrode with material during potentiodynamic cycling as the compound was not fully dissolved in DCM, but rather seemed to form a dispersion. A measurement was conducted in DMF, but the compound also formed a dispersion in this solvent, and the obtained CV data was even less indicative as the current intensities were very low, especially for the oxidation (see chapter 8.1, Figure 58). For **2PXZ-BP-COPh** (**43e**) an additional signal was exhibited in the cathodic trace at E_{red}^c = -0.83 V vs. SCE. This additional signal was also observed in the literature-reported CV data for **2PXZ-BP-COPh** (**43e**) by Wang and coworkers.[138] However, as Wang and coworkers reported the scan at positive potentials in DCM but the scan at negative potentials in DMF, a direct comparison of the literature values to the experimentally obtained values in this work is difficult. They obtained a HOMO level of -5.49 eV (DCM) and a LUMO level of -3.24 eV (DMF) as compared to the obtained values of -5.21 eV (DCM) and -3.35 eV (DCM), respectively. The discrepancy between the HOMO values arises as Wang and coworkers calculated the HOMO energy from the onset of the reduction potential, whereas in this work, the reduction potential is obtained from the following formula: $E_{red} = (E_c + E_a)/2$.

*

Finally, the 2DMAC-BP-R series was investigated by employing cyclic voltammetry and the results are shown in Figure 26 and Table 5, entries 10-12.

The oxidation potentials of **2DMAC-BP-2Me** (**44d**) and **2DMAC-BP-COPh** (**44e**) were determined as E_{ox} = 1.07 and 0.97 V vs. SCE, respectively. Based on the oxidation potential, the HOMO energy levels amount to -5.41 and -5.31 eV for **2DMAC-BP-2Me** (**44d**) and **2DMAC-BP-COPh** (**44e**), respectively. In a characteristic fashion for the DMAC donor, the scans at positive potentials of **2DMAC-BP-2Me** (**44d**) and **2DMAC-BP-COPh** (**44e**) show one other reversible process at 0.77 and 0.65 V vs. SCE, respectively, which was observed for **2DMAC-BP-F** (**39b**) (Figure 23) as well. Furthermore, **2DMAC-BP-2Me** (**44d**) shows other signals in the cathodic trace at -0.05 V vs. SCE and a hinted peak in the anodic trace at 1.09 V vs. SCE. **2DMAC-BP-COPh** (**44e**) also depicts these two additional signals at 0.05 and 1.01 V vs. SCE. The reduction potential was observed at -1.38 and -1.22 V vs. SCE for **2DMAC-BP-**

2Me (44d) and **2DMAC-BP-COPh (44e)**. Thus, the LUMO energy levels were calculated to -2.96 and -3.12 eV, respectively. Following these findings, the HOMO-LUMO gap equals 2.45 and 2.19 eV, respectively.

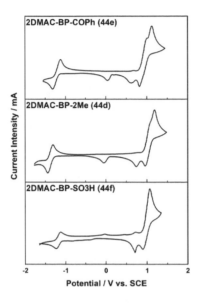

Figure 26. Cyclic voltammograms of the **2DMAC-BP-COPh (44e)** and **2DMAC-BP-2Me (44d)** measured in Ar-saturated DCM, as well as **2DMAC-BP-SO₃H (44f)** measured in Ar-saturated DMF containing 0.1 M ⌊nBu₄N⌋⌊PF₆⌋ at a scan rate of 100 mV s⁻¹.

In comparison, the HOMO level of the benzoyl-substituted **2DMAC-BP-COPh (44e)** is shallower than for the dimethyl-substituted **2DMAC-BP-2Me (44d)**. Furthermore, the benzoyl substituent stabilizes the LUMO, which indicates that this substituent act as an electron-withdrawing auxiliary acceptor. This follows the DFT calculation results reported by Zhao and coworkers, who showed that the LUMO spreads onto the carbonyl group.[139] They also showed CV results, but as the reductive scan was measured in DMF, and both the reductive and the oxidative scan were obtained at a different scan rate of 50 mV s⁻¹, a direct comparison is not feasible. However, the reported HOMO and LUMO levels of -5.28 eV (DCM) and -3.46 eV (DMF), respectively,[139] lie in a similar range as the abovementioned experimentally obtained values of -5.31 eV (DCM) and -3.12 eV (DCM).

Furthermore, **2DMAC-BP-SO₃H** (**44f**) was measured in Ar-saturated DMF due to poor solubility in DCM and can therefore not be directly compared to its fellow 2DMAC-BP-R-type compounds. The oxidation potential amounts to $E_{ox} = 0.99$ V vs. SCE. The HOMO energy level was therefore calculated as -5.33 eV. In the voltammogram, an additional signal was found at 0.71 V vs. SCE, as well as a minor signal in the anodic trace at -0.03 V vs. SCE. The reduction potential is reversible and amounts to $E_{red} = -1.18$ V vs. SCE and the LUMO was thus calculated to -3.16 eV. The resulting HOMO-LUMO energy gap is determined as 2.16 eV. Although a direct comparison to the results from Zhao and co-workers[139] is not possible due to measurement at different scan rates, the LUMO of the sulfonic acid-bearing **2DMAC-BP-SO₃H** (**44f**) might be less stabilized than the benzoyl-bearing analog.

*

The following section compares the data obtained from the TD-DFT calculation of the 2D-BP-D, 2D-BP-F and 2D-BP-O*t*Bu series. The results are summarized in Table 4.

Table 4. Summary of the TD-DFT calculation results provided by Changfeng Si (Zysman-Colman Group, University of St. Andrews) for the 2D-BP-D, 2D-BP-F and 2D-BP-O*t*Bu series; computation in the gas phase using the Pbe1pbe/6-31G(d,p) functional.

Entry	Compound	HOMO [eV]	HOMO-1 [eV]	HOMO-2 [eV]	HOMO-4 [eV]	LUMO [eV]	$\Delta E_{HOMO\text{-}LUMO}$ [eV]	S_1 [eV]	T_1 [eV]	f	ΔE_{ST} [eV]
1	2PXZ-BP-DTCz (38)	-4.95	n/a	n/a	n/a	-2.36	2.23	1.83	1.82	0.017	0.01
2	2DTCz-BP-DTCz (39d)	-5.41	n/a	n/a	n/a	-2.42	2.99	2.30	2.10	0.0401	0.20
3	2PXZ-BP-F (39a)	-4.97	-4.99	n/a	n/a	-2.58	2.39	1.89	1.87	0.005	0.02
4	2DMAC-BP-F (39b)	-5.10	-5.20	n/a	n/a	-2.51	2.67	2.16	2.15	0	0.01
5	2DTCz-BP-F (39c)	-5.48	n/a	n/a	n/a	-2.31	3.17	2.73	2.37	0.167	0.36
6	2PXZ-BP-O*t*Bu (40a)	-4.90	-4.91	n/a	n/a	-2.28	2.62	2.11	2.10	0.0015	0.01
7	2DMAC-BP-O*t*Bu (40b)	-5.10	n/a	-6.07	n/a	-2.19	2.91	2.39	2.35	0	0.04
8	2DTCz-BP-O*t*Bu (40c)	-5.48	-5.42	n/a	-6.02	-2.07	3.31	2.85	2.31	0.1209	0.53

In comparing the three donors, PXZ shows the shallowest HOMO level in all three molecule series, thereby confirming the expectation as the strongest donor. Furthermore, the values of the HOMO energy for the following comparison groups lie in a close range regardless of the nature of the varying substituent: (i) **2PXZ-BP-DTCz** (**38**), **2PXZ-BP-F** (**39a**), and **2PXZ-BP-O*t*Bu** (**40a**), (ii) **2DMAC-BP-F** (**39b**) and **2DMAC-BP-O*t*Bu** (**40b**), and (iii) **2DTCz-BP-DTCz** (**39d**), **2DTCz-BP-F** (**39c**), and **2DTCz-BP-O*t*Bu** (**40c**).

As the DFT calculations affirmed 10*H*-phenoxazine as a strong donor, the LUMO levels of **2PXZ-BP-DTCz (38)**, **2PXZ-BP-F (39a)**, and **2PXZ-BP-O*t*Bu (40a)** were compared next. The results show that the LUMO levels are increasingly stabilized in the following substituent order: O*t*Bu > DTCz > F. The HOMO-LUMO gap follows a different trend and increases from **2PXZ-BP-DTCz (38)**, which shows the smallest energy gap, over **2PXZ-BP-F (39a)** to **2PXZ-BP-O*t*Bu (40a)**, which displays the highest energy difference between the frontier molecular orbitals. When comparing the excited state energies, **2PXZ-BP-DTCz (38)** displays the lowest S_1 state, followed by **2PXZ-BP-F (39a)** and **2PXZ-BP-O*t*Bu (40a)**. This is in favor of the initial strategy to red-shift the emission by adding more donors. However, as mentioned before, this might also increase the de-excitation *via* non-radiative pathways. The same trend is observed for the T_1 state energies. Regarding the oscillator strengths, **2PXZ-BP-O*t*Bu (40a)** depicts an almost ten-time decrease in comparison to **2PXZ-BP-DTCz (38)**. The energy difference ΔE_{ST} is very small for all three compounds and hints at efficient RISC and excellent TADF potential.

When comparing the calculated data of **2DMAC-BP-O*t*Bu (40b)** to **2DMAC-BP-F (39b)**, it is observed that the fluorine substituent considerably stabilizes the LUMO. Furthermore, the S_1 state is energetically lowered. Thereby, from a computational point of view, **2DMAC-BP-F (39b)** is anticipated to show a red-shifted emission in comparison to **2DMAC-BP-O*t*Bu (40b)**. This aligns with the argumentation of fluorine as an auxiliary acceptor, while the *tert*-butoxy substituent is known as an electron-donating group that destabilizes the LUMO of the acceptor core. Both compounds show a minuscule ΔE_{ST}, which is favorable for TADF emitters.

Lastly, the compounds with DTCz donors (entry 2,5 and 8) show the lowest LUMO for **2DTCz-BP-DTCz (39d)**, followed by **2DTCz-BP-F (39c)**, and the highest LUMO value is observed for **2DTCz-BP-O*t*Bu (40c)**. The same trend is exhibited for the S_1 energies. The oscillator strength is the highest for **2DTCz-BP-F (39c)** among all the compared compounds and implicates bright fluorescence. Furthermore, the ΔE_{ST} is high for all three compounds and indicates that efficient RISC is aggravated.

Especially the comparison between the 2D-BP-D and 2D-BP-F compounds shows that a leaner molecule design with decreased molecular weight does not necessarily compromise the qualities necessary for efficient TADF and that auxiliary acceptors depict a promising molecule design strategy.

*

The data obtained from the cyclic voltammetry measurements are also compared and discussed in the following section. A summary thereof is displayed in Table 5. First of all, the respective compounds with the same donor moieties (e.g., PXZ, DMAC, or DTCz) show reduction potentials and HOMO energies in a very narrow range, which hints at a small electronic coupling between the respective acceptor and donor moieties. However, the more interesting comparison is between the LUMO energy levels that indicate the acceptor strength.

Table 5. Summary of the data obtained from cyclic voltammetry measurements of the four series 2D-BP-D (entry 1-2), 2D-BP-F (entry 3-5), 2D-BP-O*t*Bu (entry 6) and 2D-BP-R (entry 7-12) measured in Ar-saturated DCM (or DMF) containing 0.1 M [nBu$_4$N][PF$_6$] at a scan rate of 100 mV s^{-1}.

Entry	Compound	E_{ox} [V vs. SCE]	E_{red} [V vs. SCE]	E_{HOMO} [eV]	E_{LUMO} [eV]	$\Delta E_{HOMO-LUMO}$ [eV]
1	**2PXZ-BP-DTCz (38)***	0.80	-1.08	-5.14	-3.26	1.89
2	**2DTCz-BP-DTCz (39d)***	1.29	-1.11	-5.63	-3.23	2.40
3	**2PXZ-BP-F (39a)**	0.80	-1.21	-5.14	-3.13	2.01
4	**2DMAC-BP-F (39b)**	1.00	-1.19	-5.34	-3.15	2.20
5	**2DTCz-BP-F (39c)**	1.32	-1.18	-5.66	-3.16	2.50
6	**2PXZ-BP-O*t*Bu (40a)**	0.83	-1.25	-5.17	-3.09	2.07
7	**2PXZ-BP-2F (43c)**	0.85	-1.15	-5.19	-3.19	2.00
8	**2PXZ-BP-2Me (43d)**	0.80	-1.36	-5.14	-2.98	2.16
9	**2PXZ-BP-COPh (43e)**	0.87	-0.99	-5.21	-3.35	1.86
10	**2DMAC-BP-2Me (40b)**	1.07	-1.38	-5.41	-2.96	2.45
11	**2DMAC-BP-COPh (44e)**	0.97	-1.22	-5.31	-3.12	2.19
12	**2DMAC-BP-SO$_3$H (44f)***	0.99	-1.18	-5.33	-3.16	2.16

* Measured in DMF.

When comparing **2PXZ-BP-F (39a)** (entry 3) to **2PXZ-BP-O*t*Bu (40a)** (entry 6), it can be observed that the LUMO was deeper for the fluorinated compound by 0.04 eV. Although this is lower than anticipated by means of DFT calculation, it follows the predicted trend. Furthermore, the effect of auxiliary acceptor such as fluorine seems to be additive, as **2PXZ-BP-2F (43c)** (entry 7) shows a deeper LUMO energy, whereas the difference in the LUMO of the monofluorinated compound amounts to 0.06 eV. The LUMO is increasingly stabilized in the following substituent order: 2Me > O*t*Bu > F > 2F > COPh. Among the monosubstituted compounds, the benzoyl substituent deepens the LUMO significantly in comparison to the fluorine substituent by 0.22 eV. Moreover, the HOMO-LUMO gap decreases in the same substituent order as the LUMO levels.

Next, the LUMO levels of the DMAC-based emitters (entries 4, 10-11) are compared. The LUMO energy was stabilized in the following substituent order: 2Me > COPh > F. In contrast to the PXZ-based emitters, the LUMO level of **2DMAC-BP-F (39b)** is slightly deeper than the LUMO level of **2DMAC-BP-COPh (44e)**. However, their difference is comparatively small and amounts to 0.03 eV. The HOMO-LUMO gap for these compounds is very similar as they were determined as 2.20 and 2.19 eV for **2DMAC-BP-F (39a)** and **2DMAC-BP-COPh (44e)**, respectively. As the sulfonic acid-bearing compound (entry 12) was not measured in DCM, it was not included in this comparison. For the same reason, **2PXZ-BP-DTCz (38)** and **2DTCz-BP-DTCz (39d)** were not compared.

*

The synthesized 2D-BP-D, 2D-BP-F, 2D-BP-OtBu, and 2D-BP-R-type molecules were computed by TD-DFT and investigated through cyclic voltammetry. The results thereof show that the trend predicted by the TD-DFT calculations is observed in the CV measurements. Regarding the effect of the substituents and numbers of substituents it can be concluded that electron-withdrawing substituents and the amount thereof can stabilize the LUMO level and increase the acceptor strength. Furthermore, these results show that the acceptor strength of the dibenzo[a,c]phenazine acceptor core can easily be tuned by choosing an appropriate substitution pattern. The tuning of desired properties can moreover be achieved by donor selection, as both the TD-DFT calculations and CV experiments demonstrated the influence of donor strength.

76

2D-BP-D and 2D-BP-R-Type Structures – Absorption and Emission

Following the analysis of the DFT calculation results and CV measurements, the 2D-BP-D- and 2D-BP-R-type compounds were investigated using UV-vis and fluorescence spectroscopy to gain data on their absorption and emission properties. This subchapter presents the data obtained for the 2D-BP-D, 2D-BP-F, 2D-B-O*t*Bu, 2PXZ-BP-R, and 2DMAC-BP-R. At the end of this subchapter, the summary of the obtained results is displayed in Table 8.

*

First, the 2D-BP-D series was investigated through UV-vis and fluorescence spectroscopy in comparison to the dibrominated precursor **2Br-BP-DTCz (37c)**. The UV-vis measurements were carried out by Changfeng Si (University of St. Andrews, Zysman-Colman group), and the results thereof are shown together with the data obtained from fluorescence spectroscopy measurements in Figure 27.

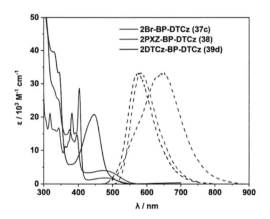

Figure 27. UV-vis and fluorescence spectroscopy spectra of the 2D-BP-D series comprising **2PXZ-BP-DTCz (38)** and **2DTCz-BP-DTCz (39d)** and the precursor **2Br-BP-DTCz (37c)** measured in toluene. The UV-vis data was provided by Changfeng Si (University of St. Andrews, Zysman-Colman group). The emission data was obtained from 0.1mM toluene solutions and is depicted as normalized values in dotted lines.

The UV-vis data was visualized as a plot of the molar extinction coefficient against the wavelength. Absorption maxima for the precursor **2Br-BP-DTCz (37c)** are observed at 318, 381, 403, and 484 nm. The tri-donor-substituted **2PXZ-BP-DTCz (38)** shows absorption peaks at 342, 375, 395, and 476 nm. Zhao and co-workers had reported that their tri-PXZ-substituted

3PXZ-BP (3c) displays absorption peaks around 280 nm, which can be ascribed to π→π* transition of the conjugated skeleton, while the long-wavelength absorption bands between 410-600 nm originate from ICT from the PXZ donor to the BP acceptor.[42] Based on this report, the absorption bands at 484 and 476 were assigned to ICT transitions for **2Br-BP-DTCz (37c)** and **2PXZ-BP-DTCz (38)**.

On the other hand, **2DTCz-BP-DTCz (39d)** showed absorption maxima, and the respective extinction coefficients in logarithmic values in parenthesis, at 331 (4.60), 344 (4.53), and 447 nm (4.32 log(M^{-1} cm^{-1})). Singh *et al.* reported a comparable tri-carbazole-substituted BP derivative.[142] Although the molecular structures of the two compounds, as shown in Figure 21 exhibit different substitution patterns to the BP core, they are structurally related. They reported absorption peaks for **Dye 4 (45)** at 286 (5.11), 339 (4.51), 385 (4.33), 406 (4.50), and 453 nm (4.38 log(M^{-1} cm^{-1})). The absorption peak at 286 nm was assigned to a localized π→π* transition of the carbazole moiety, while the weaker signals between 330-385 nm were identified as n→π* or π→π* transitions that originate from orbitals delocalized over the conjugated system. Lastly, the signals above 400 nm were ascribed to ICT transitions from the donor to the acceptor.[142] Based on this report, the absorption peak of **2DTCz-BP-DTCz (39d)** at 447 nm was also attributed to the ICT from the DTCz donors to the BP acceptor. Compared to **Dye 4 (45)**, **2DTCz-BP-DTCz (39d)** depicts only one absorption maximum over 400 nm with a hypsochromic shift by 6 nm. The extinction coefficients for both compounds for the longest absorption wavelength are almost identical. Following this observation, it can be assumed that the absorption of light through an ICT transition is equally strong for **2DTCz-BP-DTCz (39d)** and **Dye 4 (45)**.

Compared to **2Br-BP-DTCz (37c)** and **2PXZ-BP-DTCz (38)**, **2DTCz-BP-DTCz (39d)** shows a different absorption spectrum. The geometry optimization of the DFT calculation had shown that the HOMO is mainly located on the DTCz donor introduced through S_NAr. Thus, the nature of the ICT transition is directed by different parts of the molecules, which shows in the absorption spectrum. Furthermore, the absorption through the ICT state is strongest for **2DTCz-BP-DTCz (39d)** (Table 8) as it shows a five-fold increase in the extinction coefficient in comparison to **2PXZ-BP-DTCz (38)**. An explanation for this observation is found when looking at the oscillator strength, as the extinction coefficient and the oscillator strength are related by equation (4) (chapter 1.1). The calculated oscillator strength for **2DTCz-BP-DTCz (39d)** is higher than for **2PXZ-BP-DTCz (38)** (Table 4).

Furthermore, the maximum emission wavelengths of the DTCz-only compounds are in a similar range as **2Br-BP-DTCz (37c)** shows a maximum at 586 nm and **2DTCz-BP-DTCz (39d)** at 574 nm. **2PXZ-BP-DTCz (38)** on the other hand, depicts a red-shifted emission at 647 nm. These results indicate that the stronger PXZ donor shifts the emission maximum of **2PXZ-BP-DTCz (38)** to longer wavelengths than **2DTCz-BP-DTCz (39d)**.

*

The photophysical characterization and analysis of the obtained data of the 2D-BP-F series were entirely carried out by Changfeng Si (University of St. Andrews, Zysman-Colman group) with the help of Dr. Tomas Matulaitis regarding the ΔE_{ST} measurements. Subsequently, this molecule series was employed in OLED devices by Dr. Abhishek Kumar Gupta (University of St. Andrews, Zysman-Colman group, and Organic Semiconductor Optoelectronics group). The results were successfully published, and as they are original to the respective contributions from collaborators, the results are presented as an excerpt in this work (Table 6 and Table 7), while the interested reader is referred to the paper for more detail.[40]

The UV-vis absorption spectra of **2PXZ-BP-F (39a)**, **2DMAC-BP-F (39b)**, and **2DTCz-BP-F (39c)** were measured in degassed toluene solution prepared by three freeze–pump–thaw cycles and show strong absorption bands around 310 nm, which were attributed to locally excited (LE) $\pi{\rightarrow}\pi^*$ transitions of the donors and BP-F units.[29, 106, 120] The weaker and broader absorption bands between 410-520 nm, on the other hand, were attributed to ICT transitions from the electron-donating to the electron-accepting units.[105] When comparing the ICT bands, it was observed that **2DTCz-BP-F (39c)** ($\lambda_{abs} = 440$ nm, $21{\cdot}10^3$ M^{-1} cm^{-1}) had a higher extinction coefficient than **2DMAC-BP-F (39b)** ($\lambda_{abs} = 450$ nm, $3{\cdot}10^3$ M^{-1} cm^{-1}) and **2PXZ-BP-F (39a)** ($\lambda_{abs} = 476$ nm, $3{\cdot}10^3$ M^{-1} cm^{-1}). As shown by DFT calculation, the DTCz groups are less twisted, which leads to an increased conjugation and oscillator strength. Furthermore, as expected, the ICT band was shifted to lower energies with increasing donor strength.[106] The photoluminescence spectra showed unstructured and broad signals, which suggest an excited state with a strong ICT character and the maximum emission wavelength for **2PXZ-BP-F (39a)**, **2DMAC-BP-F (39b)**, and **2DTCz-BP-F (39c)** were observed at 674, 589 and 505 nm, respectively. The CT nature of the emissive excited state was confirmed by positive solvatochromism for all three compounds. Moreover, photoluminescence quantum yields Φ_{PL} in degassed toluene solution were obtained as 8, 30, and 51% for **2PXZ-BP-F (39a)**,

2DMAC-BP-F (**39b**), and **2DTCz-BP-F** (**39c**), respectively. Upon exposure to oxygen, a decrease thereof to 6, 21, and 49% was observed.

Table 6. Summary of the photophysical investigation of the 2D-BP-F series (**39a-c**) conducted by Changfeng Si (University of St. Andrews, Zysman-Colman group). The absorption and emission wavelengths, as well as the prompt and delayed lifetimes, and the photoluminescence quantum yields in solution were measured in degassed toluene prepared by three freeze–pump–thaw cycles. The ΔE_{ST} values in solution were calculated from measurements in 2-MeTHF at 77 K unless otherwise indicated.

Entry	Compound	λ_{em} [nm]	τ_p [ns]	τ_d [ms]	ΔE_{ST} [eV]	Φ_{PL} [%]	k_{ISC} [10^7 s^{-1}]	k_{RISC} [10^5 s^{-1}]
	in solution							
1	**2PXZ-BP-F**	647	16.9	0.2	0.05*	8 (6)	n/a	n/a
2	**2DMAC-BP-F**	589	27.7	19.0	0.20	30 (21)	n/a	n/a
3	**2DTCz-BP-F**	505	6.6	n/a	0.43	51 (49)	n/a	n/a
	in PMMA							
4	**2PXZ-BP-F** (5 wt%)	615	n/a	2.99	0.06	28 (17)	n/a	n/a
5	**2DMAC-BP-F** (10 wt%)	588	n/a	5033	0.28	48 (34)	n/a	n/a
6	**2DTCz-BP-F** (1.5 wt%)	524	n/a	10105	0.34	29 (27)	n/a	n/a
	in mCBP							
7	**2PXZ-BP-F** (5 wt%)	611	31.0	1.83	0.02	58 (47)	1.52	2.41
8	**2DMAC-BP-F** (10 wt%)	584	19.6	90.6	0.11	78 (48)	2.45	0.133
9	**2DTCz-BP-F** (1.5 wt%)	522	4.3	10152	0.30	60 (47)	10.9	0.000514

* Measured in hexane.

Experimental values for the ΔE_{ST} of **2DMAC-BP-F** (**39b**) and **2DTCz-BP-F** (**39c**) amount to 0.20 and 0.43 eV, respectively, and were obtained from prompt fluorescence and phosphorescence spectra in 2-MeTHF. The ΔE_{ST} of **2PXZ-BP-F** (**39a**) was obtained from prompt fluorescence and phosphorescence spectra in hexane and amounts to 0.05 eV. The photoluminescence decays were monitored using time-correlated single-photon counting (TCSPC), and the ICT band of **2PXZ-BP-F** (**39a**) and **2DMAC-BP-F** (**39b**) decays with prompt fluorescence lifetimes, τ_p, of 16.9 and 27.7 ns, while the delayed fluorescence lifetimes, τ_d, amount to 0.2 and 19.0 μs, respectively. The accessibility of the triplet states follows from the observation that the delayed emission is strongly quenched upon oxygen exposure. In contrast to these two compounds, **2DTCz-BP-F** (**39c**) showed a prompt fluorescence lifetime of 6.6 ns and no delayed emission.

In addition to the investigation of the emission in solution, the solid-state photoluminescence was examined in doped PMMA films and mCBP, as the latter is an OLED host matrix with sufficiently high triplet energy (T_1 = 2.84 eV) to confine the excitons onto the emitter.[143] The results thereof are shown in Table 6. The emission wavelengths of **2PXZ-BP-F** (**39a**), **2DMAC-BP-F** (**39b**), and **2DTCz-BP-F** (**39c**) were determined as 615, 588, and 524 nm in doped PMMA films, respectively. Furthermore, the PLQYs under N_2 atmosphere amount to 28, 48, and 29%, respectively. The photoluminescence quantum yield decreased upon oxygen exposure. Furthermore, the average delayed lifetimes were obtained as 2.99 μs, 5.03 ms, as well as 10.1 ms for **2PXZ-BP-F** (**39a**), **2DMAC-BP-F** (**39b**), and **2DTCz-BP-F** (**39c**), respectively. Therefore, thermally activated delayed fluorescence is effective for the 2D-BP-F series in doped PMMA films. The estimated ΔE_{ST} amounts to 0.06, 0.28 and 0.34 eV for **2PXZ-BP-F** (**39a**), **2DMAC-BP-F** (**39b**), and **2DTCz-BP-F** (**39c**), respectively.

The investigation of the emitters doped into the mCBP host showed emission maxima at 611, 584, and 522 nm for **2PXZ-BP-F** (**39a**), **2DMAC-BP-F** (**39b**), and **2DTCz-BP-F** (**39c**), respectively. The PLQYs under N_2 atmosphere were higher than in doped PMMA and amounted to 58, 78, and 60%, respectively, and decreased upon oxygen exposure. As observed in doped PMMA films, the measurements in the doped mCBP host also showed the highest delayed lifetime for **2DTCz-BP-F** (**39c**) (10.2 ms), followed by **2DMAC-BP-F** (**39b**) (90.6 μs), and **2PXZ-BP-F** (**39a**) (1.83 μs), and indicated TADF. The ΔE_{ST} in doped mCBP films was estimated to amount to 0.02, 0.11, and 0.30 eV for **2PXZ-BP-F** (**39a**), **2DMAC-BP-F** (**39b**), and **2DTCz-BP-F** (**39c**), respectively. Furthermore, the corresponding rate constants of ISC and RISC are depicted in Table 6 and show that while **2PXZ-BP-F** (**39a**) shows the slowest intersystem crossing rate, the reverse intersystem crossing is significantly faster than for **2DMAC-BP-F** (**39b**) and **2DTCz-BP-F** (**39c**). Moreover, the TADF character of these three emitters in doped mCBP films was confirmed as the relative intensities of the delayed photoluminescence increased with increasing temperature from 100 K to 300 K. However, **2DTCz-BP-F** (**39c**) was found to show only inefficient TADF, which was explained by the large ΔE_{ST} that aggravates RISC, which was observed in the k_{RISC} value as well.

Subsequent to the photophysical characterization, Dr. Abhishek Kumar Gupta (University of St. Andrews, Zysman-Colman group, and Organic Semiconductor Optoelectronics group) investigated the emitters in OLED devices, and the results thereof are summarized in Table 7.

Table 7. Summary of the emitter performance in OLED devices (ITO/HATCN (5 nm)/NPB (40 nm)/TCTA (10 nm)/emissive layer (20 nm)/TmPyPB (40 nm)/LiF (0.6 nm)/Al (100 nm)) built by Dr. Abhishek Kumar Gupta (University of St. Andrews, Zysman-Colman group and Organic Semiconductor Optoelectronics group).

Entry	Compound	$\lambda_{EL}{}^a$	$V_{on}{}^b$	CE^c	PE_{max}	EQE^c	CIE^a
		[nm]	[V]	[cd A^{-1}]	[lm W^{-1}]	[%]	[x,y]
1	**2PXZ-BP-F** (5 wt%)	605	3.3	26.3/19.1/12.7	23.0	12.4/9.3/6.3	0.549, 0.444
2	**2DMAC-BP-F** (10 wt%)	585	3.6	59.7/23.5/8.1	55.4	21.8/8.7/3.3	0.513, 0.479
3	**2DTCz-BP-F** (1.5 wt%)	518	3.7	6.7/3.7/3.0	5.6	2.1/1.2/1.0	0.290, 0.580

[a] The electroluminescence maximum and CIE coordinates recorded at 5 V. [b] Voltage at a brightness of 1 cd m^{-2}. [c] The order of measured values: the maximum EQE / EQE at 100 cd m^{-2} / EQE at 1000 cd m^{-2}.

The OLED devices were fabricated by vacuum deposition and displayed the following OLED device architecture: indium tin oxide (ITO)/1,4,5,8,9,11-hexaazatriphenylene- hexacarbonitrile (HATCN) (5 nm)/N,N'-di(1-naphthyl)- N,N'-diphenyl-(1,1'-biphenyl)-4,4'-diamine (NPB) (40 nm)/tris(4-carbazoyl-9-ylphenyl)amine (TCTA) (10 nm)/emissive layer (20 nm)/1,3,5-tri[(3-pyridyl)-phen-3-yl]benzene (TmPyPB) (40 nm)/ LiF (0.6 nm)/Al (100 nm). Hereby, HATCN, NPB, and TCTA represent the HIL, HTL, and electron blocking layer (EBL), respectively. In the OLED device architecture, TmPypPB plays the role of both the ETL and hole blocking layer (HBL), while LiF acts as an EIL. The EML of the three devices contained 1.5 wt% **2PXZ-BP-F** (**39a**), 10 wt% **2DMAC-BP-F** (**39b**), or 5 wt% **2DTCz- BP-F** (**39c**) doped into mCBP, respectively. All three emitters showed turn-on voltages of their respective devices in a similar range between 3.3-3.7 V. The turn-on voltage is dependent on the energy difference between the HOMOs of the materials used in the HTL and EML. The longest electroluminescence wavelength was observed for **2PXZ-BP-F** (**39a**) at 605 nm, and a donor strength-dependent trend is also observed here. The CIE coordinates were determined as (0.55, 0.44), (0.51, 0.48) and (0.29, 0.58) for the devices featuring with **2PXZ-BP-F** (**39a**), **2DMAC-BP-F** (**39b**) and **2DTCz-BP-F** (**39c**), respectively. The highest maximum external quantum efficiency (EQE) of 21.8% was observed for the device with **2DMAC-BP-F** (**39b**), which also showed the highest maximum current efficiency (CE$_{max}$) of 59.7 cd A^{-1} and maximum power efficiency (PE$_{max}$) of 55.4 lm W^{-1}. However, the efficiency roll-off of the **2DMAC-BP-F** (**39b**)-employing device was quite stark, as the maximum EQE dropped from 21.8% to 3.3% at the highest brightness of 1000 cd m^{-2}. In contrast, the **2PXZ-BP-F** (**39a**)-based device showed a lower maximum EQE of 12.4% but a moderate roll-off efficiency with an EQE of 6.3% at 1000 cd m^{-2}. The better efficiency roll-off performance of **2PXZ-BP-F** (**39a**) in comparison to **2DMAC-BP-F (39b)** was reasoned to partly be because of the relatively short τ_d. A short delayed lifetime means that there will be a lower triplet exciton concentration, which leads to a lower probability of undesired non-radiative exciton quenching.[144-145] **2DTCz-BP-F (39c)** had shown a high ΔE_{ST} in the DFT calculations and experimental estimation. Therefore, it was not surprising to observe the lowest EQE_{max} of 2.1% as the triplet excitons could not be efficiently harvested and the long delayed lifetimes allowed non-radiative exciton quenching through triplet-triplet annihilation and triplet-polaron annihilation.

The 2D-BP-F series was extensively investigated in cooperation with Changfeng Si and Dr. Abhishek Kumar Gupta from the Zysman-Colman group at the University of St. Andrews regarding its photophysical properties in solution, doped into PMMA and mCBP film and in OLED devices. The emitters demonstrated TADF properties according to the calculated and estimated ΔE_{ST}, as well as good photoluminescence quantum yield in doped mCBP film. Furthermore, the OLED device featuring **2DMAC-BP-F (39b)** showed an excellent EQE_{max} of 21.8%, while the **2PXZ-BP-F (39a)**-employing device displayed better roll-off efficiency.

*

Next, the absorption and emission of the 2D-BP-O*t*Bu compounds were investigated, and the results thereof are displayed in Figure 28. **2PXZ-BP-O*t*Bu (40a)** shows absorption peaks at 311, 372, 392, and 472 nm, while **2DMAC-BP-O*t*Bu (40b)** exhibits signals at 373, 394, and 438 nm. In contrast to these two compounds, **2DTCz-BP-O*t*Bu (40c)** shows absorption peaks at 346 and 433 nm. Following the observation and literature comparison of the previous compound series discussed in this subchapter, the absorption bands above 400 nm were assigned to ICT transitions. While **2PXZ-BP-O*t*Bu (40a)** and **2DMAC-BP-O*t*Bu (40b)** indicate broad bands originating from ICT with low extinction coefficients of 3.01 and $2.60 \cdot 10^3$ M^{-1} cm^{-1}, **2DTCz-BP-O*t*Bu (40c)** demonstrates a narrower peak with an immensely increased extinction coefficient of $26.14 \cdot 10^3$ M^{-1} cm^{-1} (Figure 28b). The extinction coefficient trend aligns well with the calculated oscillator strength summarized in Table 4. Out of the three 2D-BP-O*t*Bu-type compounds, the **2PXZ-BP-O*t*Bu (40a)** displays the longest ICT absorption wavelength. This means that the ICT from an intramolecular electron-donating to an electron-accepting moiety requires less energy.

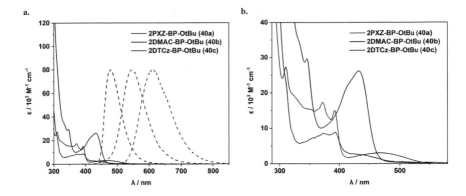

Figure 28. a. UV-vis spectra and fluorescence spectra of the 2D-BP-O*t*Bu series measured in toluene. The UV-vis data for **2DTCz-BP-OtBu (40c)** was provided by Changfeng Si (University of St. Andrews, Zysman-Colman group). The emission data was obtained from 0.1 mM toluene solutions and is depicted as normalized values in dotted lines. **b.** Excerpt from the UV-vis spectra.

Furthermore, the emission wavelength maxima for the 2D-BP-O*t*Bu series were obtained. **2PXZ-BP-OtBu (40a)** displays the longest emission wavelength with a maximum at 613 nm, followed by **2DMAC-BP-O*t*Bu (40b)**, which emits at 548 nm and **2DTCz-BP-O*t*Bu (40c)** that shows an emission maximum at 480 nm. The effect of the donor strength can thus also be observed in the 2D-BP-O*t*Bu series.

*

The UV-vis and fluorescence spectra of the 2PXZ-BP-R compounds **2PXZ-BP-2F (43c)**, **2PXZ-BP-2Me (43d)**, and **2PXZ-BP-COPh (43e)** are shown in Figure 29. The absorption maxima with the respective extinction coefficient in parenthesis of **2PXZ-BP-2F (43c)** were detected at 312 (8.21), 373 (6.00), 393 (6.11), and 475 nm ($1.05 \cdot 10^3$ M^{-1} cm^{-1}). A very similarly shaped absorption spectrum, but with higher extinction coefficients, was obtained for **2PXZ-BP-COPh (43e)**, which displayed peaks at 290 (52.63), 301 (53.78), 313 (38.09), 381(19.62), 402 (17.55) and 492 nm ($3.18 \cdot 10^3$ M^{-1} cm^{-1}). In contrast to these two compounds, the extinction coefficients for **2PXZ-BP-2Me (43d)** were considerably lower, and two peaks were observed at 397 (1.08) and 463 nm ($0.25 \cdot 10^3$ M^{-1} cm^{-1}). The highest extinction coefficient for the ICT band was observed for **2PXZ-BP-COPh (43e)**. Furthermore, the emission maximum of **2PXZ-BP-2F (43c)** was detected at 659 nm, while **2PXZ-BP-2Me (43d)** and **2PXZ-BP-COPh (43e)** showed maxima at 598 and 680 nm, respectively. The emission wavelength obtained for **2PXZ-BP-COPh** (**43e**) is red-shifted in comparison to the literature-reported emission wavelength (λ_{PL} = 642 nm) by Wang and coworkers. A key difference in the measurement setup was the concentration, as the measurement reported in this work had been carried out with a 0.1 mM toluene solution, while Wang and coworkers had conducted the measurement in 0.01 mM toluene solution at room temperature.[138]

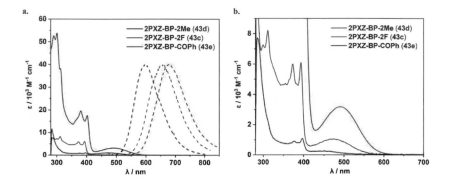

Figure 29. a. UV-vis spectra and fluorescence spectra of the 2DMAC-BP-R series measured in toluene. The emission data was obtained from 0.1 mM toluene solutions and is depicted as normalized values in dotted lines. **b.** Excerpt from the UV-vis spectra.

The effect of the acceptor substitution is distinctly portrayed in the absorption and emission spectra. The ICT bands are shifted to lower energies in the absorption spectra with an increasing electron-withdrawing effect of the attached substituent. This is in parallel to the observations made in the CV measurements, where the LUMO energies were stabilized depending on the electron-withdrawing strength of the substituent (Table 5). In the emission spectra, a similar trend is observed as **2PXZ-BP-2Me** (**43d**), **2PXZ-BP-2F** (**43c**), and **2PXZ-BP-COPh** (**43e**) show increasing wavelengths. For the 2PXZ-BP-R series, a deepening of the LUMO through stabilizing substituents occurs simultaneously to a red-shifted ICT band in the absorption spectrum and a bathochromically shifted emission wavelength.

*

Lastly, the 2DMAC-BP-R series was investigated by means of UV-vis and fluorescence spectroscopy. The UV-vis data is displayed in Figure 30 and indicates absorption peaks with respective extinction coefficients at 380 (24.27), 399 (29.59), and 435 nm ($6.85 \cdot 10^3$ M^{-1} cm^{-1}) for **2DMAC-BP-2Me** (**44d**). **2DMAC-BP-COPh** (**44e**), and **2DMAC-BP-SO₃H** (**44f**) show similar peaks at 382 (17.80) and 403 nm (18.96· 10^3 M^{-1} cm^{-1}), and 378 (11.32) and 398 nm (10.80· 10^3 M^{-1} cm^{-1}), but red-shifted ICT bands at 462 (3.48) and 465 nm (2.79· 10^3 M^{-1} cm^{-1}), respectively. The highest extinction coefficient for an ICT band was observed for **2DMAC-BP-2Me (44d)**.

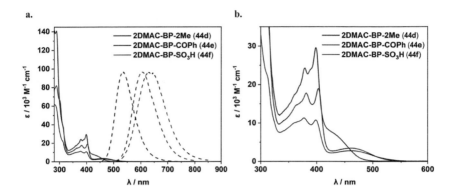

Figure 30. a. UV-vis spectra measured by the IOC Analysis Department and fluorescence spectra of the 2DMAC-BP-R series measured in toluene. The emission data was obtained from 0.1 mM toluene solutions and is depicted as normalized values in dotted lines. **b.** Excerpt from the UV-vis spectra.

The fluorescence spectra show emission maxima at 537, 605 and 637 nm for **2DMAC-BP-2Me (44d)**, **2DMAC-BP-COPh (44e)**, and **2DMAC-BP-SO3H (44f)**, respectively. Thus, the absorption and emission data in combination with the data obtained from CV measurements for the 2DMAC-BP-R series shows that electron-withdrawing substituents result in a deeper LUMO in parallel with red-shifted ICT absorption and emission wavelength.

*

Finally, the data obtained for the 2D-BP-D, 2D-BP-F, 2D-BP-O*t*Bu, and 2D-BP-R series are summarized in Table 8 and compared.

Table 8. Summary of the data obtained from the UV-vis and fluorescence spectroscopy measurements in 0.1 mM toluene solution for the 2D-BP-D, 2D-BP-F, 2D-BP-OtBu, 2PXZ-BP-R and 2DMAC-BP-R series; *Values are taken from the published data and were measured by Changfeng Si (University of St. Andrews, Zysman-Colman group) in degassed toluene solution prepared by three freeze-pump-thaw cycles; specific concentrations were not stated.[40]

Entry	Compound	λ_{max} (ε) [nm] ([10^3 M^{-1} cm^{-1}])	λ_{em} (λ_{exc}) [nm]
1	2Br-BP-DTCz (37c)	318 (21.14), 381 (21.23), 403 (28.60), 484 (1.82)	586 (484)
2	2PXZ-BP-DTCz (38)	342 (25.37), 375 (16.22), 395 (14.82), 476 (4.03)	647 (460)
3	2DTCz-BP-DTCz (39d)	331 (39.48), 344 (33.55). 447 (20.72)	573 (450)
4	2PXZ-BP-F (39a)*	310 (n/a), 476 (3)	674 (391)
5	2DMAC-BP-F (39b)*	310 (n/a), 415 (3)	589 (391)
6	2DTCz-BP-F (39c)*	310 (n/a), 440 (21)	505 (343)
7	2PXZ-BP-O*t*Bu (40a)	311 (27.26), 372 (17.30), 392 (15.01), 472 (3.01)	613 (480)
8	2DMAC-BP-O*t*Bu (40b)	373 (8.65), 394 (8.85), 438 (2.60)	548 (420)
9	2DTCz-BP-O*t*Bu (40c)	346 (29.71), 433 (26.14)	480 (390)
10	2PXZ-BP-2F (43c)	312 (8.21), 373 (6.00), 393 (6.11), 475 (1.05)	659 (450)
11	2PXZ-BP-2Me (43d)	397 (1.08), 463 (0.25)	598 (438)
12	2PXZ-BP-COPh (43e)	290 (52.63), 301 (53.78), 313 (38.09), 381(19.62), 402 (17.55), 492 (3.18)	680 (492)
13	2DMAC-BP-2Me (44d)	380 (24.27), 399 (29.59), 435 (6.85)	537 (438)
14	2DMAC-BP-COPh (44e)	382 (17.80), 403 (18.96), 462 (3.48)	605 (467)
15	2DMAC-BP-SO$_3$H (44f)	378 (11.32), 398 (10.80), 465 (2.79)	637 (467)

The comparison of the absorption peaks shows that the furthest red-shift for an absorption originating from ICT is observed for **2PXZ-BP-COPh (43e)**, followed by **2Br-BP-DTCz (37c)** and **2PXZ-BP-DTCz (38)** and **2PXZ-BP-F (39a)**. The strong red-shift of the CT state for the PXZ-based emitters in comparison to DMAC- or DTCz-based emitters is due to the donor strength of PXZ and this phenomenon was described in the literature as well.[106] A bromine substituent can in principle have three effects through which it affects: (i) inductively as an electron-withdrawing substituent, (ii) through a positive mesomeric effect or (iii) through the heavy-atom effect (HAE). The latter leads to increased spin-orbit coupling (SOC) which in turn eventually results in higher k_{RISC} and shorter delayed lifetimes.[146-147] The strong bathochromic shift in the case of the precursor molecule **2Br-BP-DTCz (37c)**, however, is assumed to stem from the electron-withdrawing effect of the bromine substituents referring to similar literature-reported observations[148] as the bromines stabilize the LUMO, which is presumably located on the BP acceptor core. Furthermore, in comparison to long-wavelength absorption of the unsubstituted **DTCz-BP** (**30c**) (λ_{abs} = 468 nm), the ICT absorption band is shifted by 16 nm, while the extinction coefficient is reduced by a factor of approximately 1.5.

The measurements of **2PXZ-BP-F** (**39a**) and **2PXZ-BP-2F** (**43c**) were carried out under different conditions and can therefore not be compared. However, the comparison with tri-donor-substituted **2PXZ-BP-DTCz** (**38**) shows that the leaner molecule design relying on an auxiliary acceptor is superior regarding the emission shift to longer wavelengths. This is also confirmed by comparing **2PXZ-BP-DTCz** (**38**) with **2PXZ-BP-COPh** (**43e**). All in all, for the 2PXZ-based emitters, the substituent effect on a bathochromic shift can be observed in the following order (from shortest to longest λ_{em}): 2Me < OtBu < DTCz < 2F < COPh. For the 2DMAC-based emitters, this order was confirmed: 2Me < OtBu < COPh < SO$_3$H. For the DTCz-based emitters, the trisubstituted **2DTCz-BP-DTCz** (**39d**) showed a longer emission wavelength than **2DTCz-BP-OtBu** (**40c**).

3.2. Pyrrolo[3,4-*f*]isoindole-1,3,5,7(2*H*,6*H*)-tetraone (PIT)-Based TADF Molecule Design

PIT - Synthesis

Building on previous work done in the Bräse group[114] on phthalimide-based TADF molecules **XI**, the pyrrolo[3,4-*f*]isoindole-1,3,5,7(2*H*,6*H*)-tetraone (PIT)-based TADF molecule design was developed and investigated. In 2014, Adachi and coworkers had reported upon fluorescent anthracene derivatives **1H (46a)** and **1 (46b)**, which had been computationally designed to suppress vibrational coupling that leads to undesirable radiationless processes such as vibrational relaxation and internal conversion, as shown in Figure 31.[149]

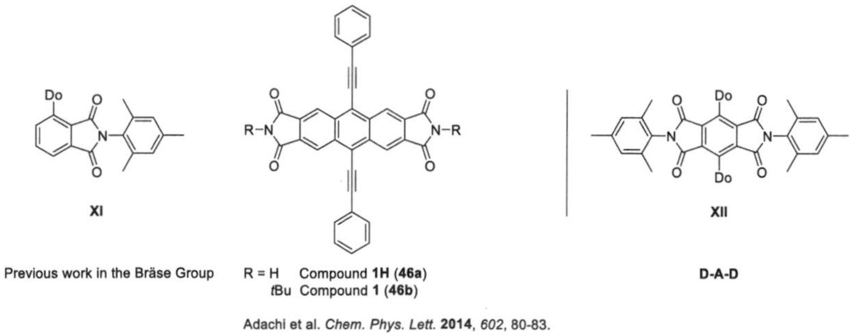

Previous work in the Bräse Group   R = H   Compound **1H (46a)**   **D-A-D**
                                          *t*Bu   Compound **1 (46b)**

Adachi et al. *Chem. Phys. Lett.* **2014**, *602*, 80-83.

Figure 31. The rationale of the PIT molecule design based on previous work done in the Bräse group[114] and reported literature by Adachi and coworkers[149].

The molecule design of the PIT scaffold resulted from a combination of the phthalimide structure and the reported anthracene derivative. The realization of the molecule design was conducted by targeting the synthesis of D-A-D-type structures **XII**.

Firstly, the PIT acceptor structure **49** was established through the imide synthesis starting from the respective anhydride **47** with 2,4,6-trimethylaniline **48a** in acetic acid, which resulted in a yield of 91% as a yellow-colorless solid as shown in Scheme 14.

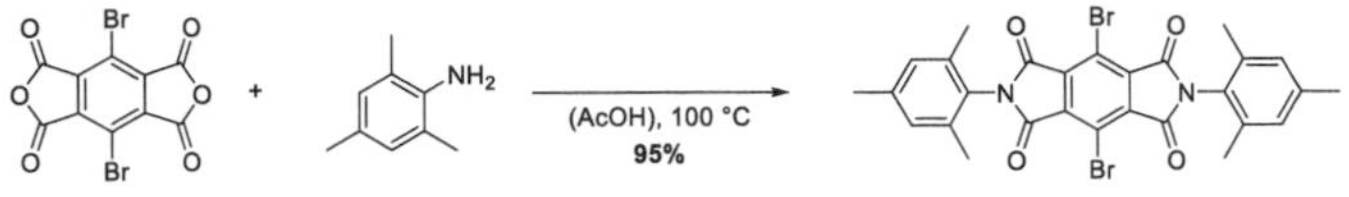

Scheme 14. Synthesis of PIT acceptor **49**.

Then, the coupling of donor units to the acceptor core *via* a BUCHWALD-HARTWIG cross-coupling was attempted. As summarized in Table 9, the synthesis of D-A-D-type compounds **50d** and **50c**, employing an acridin-9(10*H*)-one **31d** and 3,6-di-*tert*-butyl-carbazole donor **31c**, could not be realized by using these reaction conditions and catalyst systems.

Table 9. Summary of unsuccessful BUCHWALD-HARTWIG cross-coupling conditions for the synthesis of D-A-D-type structures **50c-d** based on the PIT acceptor.

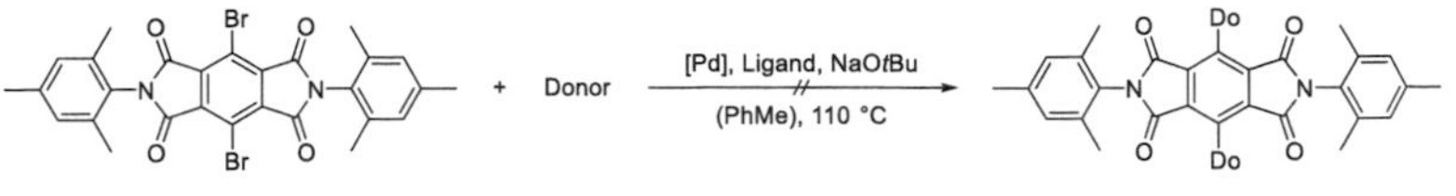

Entry	Donor	[Pd]	Ligand
1	31d	Pd$_2$(dba)$_3$	[(tBu)$_3$PH]BF$_4$
2	31c	Pd$_2$(dba)$_3$	[(tBu)$_3$PH]BF$_4$
3	31c	Pd$_2$(dba)$_3$	P(Bu)$_3$
4	31c	Pd$_2$(dba)$_3$	dppf
5	31c	Pd(OAc)$_2$	P(tBu)$_3$

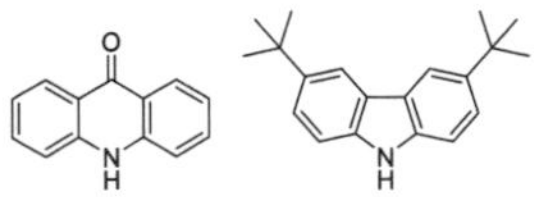

Therefore, a synthesis strategy based on a revised molecule design was investigated. A π-spacer unit was inserted in-between the D and A moieties to spatially separate the frontier molecular orbitals (FMOs), leading to D-π-A-π-D-type structures. In addition to the spatial separation and extension of the conjugated π-electron system, the π-spacer provides a wider variety of synthetic options. Here, two routes, "A" and "B", were investigated, as shown in Scheme 15. Route A is based on an initial SUZUKI cross-coupling step that attaches the π-spacer to obtain **51** and is followed by a BUCHWALD-HARTWIG coupling step to attach the donor unit. Route B is based on initially connecting the donor unit with the π-spacer to obtain **52a-d** and using the SUZUKI cross-coupling to attach it to the acceptor core **49**.

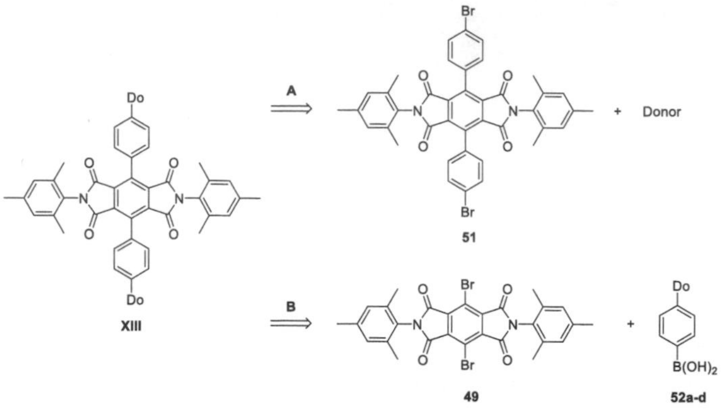

Scheme 15. Retrosynthetic strategy for the synthesis of D-π-A-π-D-type TADF emitters **XIII** based on the PIT acceptor core.

The first step of route A leads to the synthesis of the π-spacer-acceptor-conjugate **51** and is displayed in Scheme 16. Compound **51** was obtained in a yield of 59% as a yellow solid.

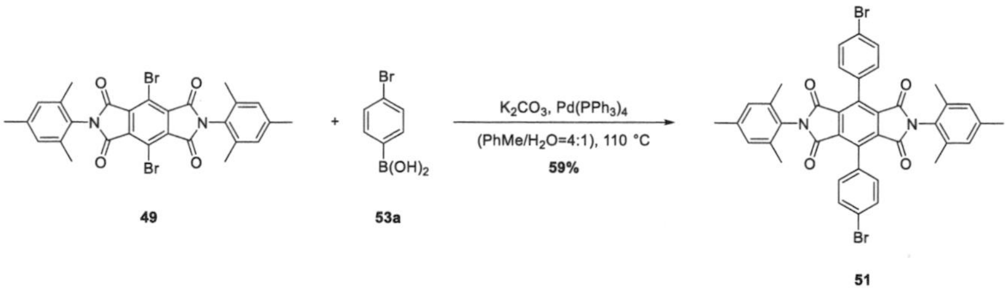

Scheme 16. Synthesis of compound **51**.

Subsequently, the BUCHWALD-HARTWIG cross-coupling with a donor unit was attempted. Here, two standard donor moieties, namely DTCz and PXZ, were used. The reaction conditions are summarized in Table 10 and were unsuccessful in synthesizing the desired conjugates **XIII**.

Table 10. BUCHWALD-HARTWIG cross-coupling conditions for the synthesis of compounds **XIII**.

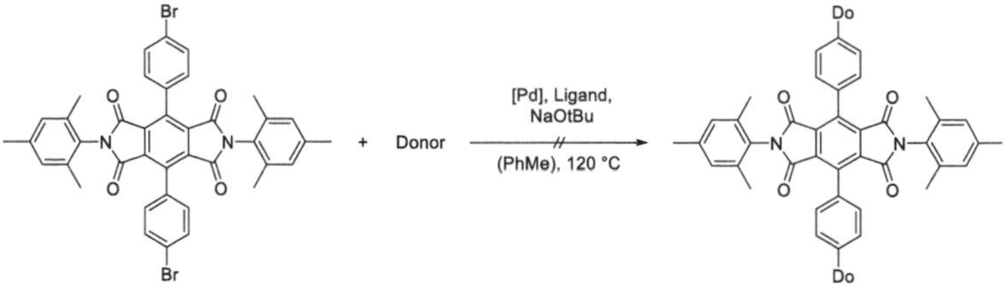

Entry	Donor	[Pd]	Ligand
1	31c	Pd(OAc)$_2$	P(tBu)$_3$
2	31a	Pd(dba)$_2$	[(tBu)$_3$PH]BF$_4$

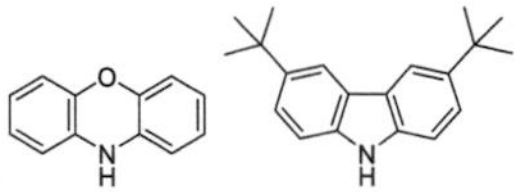

Therefore, route B was attempted. For this reaction pathway, the donor-spacer conjugates **52a-b** were prepared. The PXZ-based donor-spacer conjugate **52a** was synthesized in two steps according to literature-known procedures[10, 150] *via* ULLMANN coupling and a subsequent borylation reaction yielding the boronic acid in 36% yield over two steps as a pale-yellow solid. The yield of the first step was lower than expected because the disubstitution of substrate **54** occurs as an undesired side reaction, which led to compound **56** in 14% yield, as shown in Scheme 17.

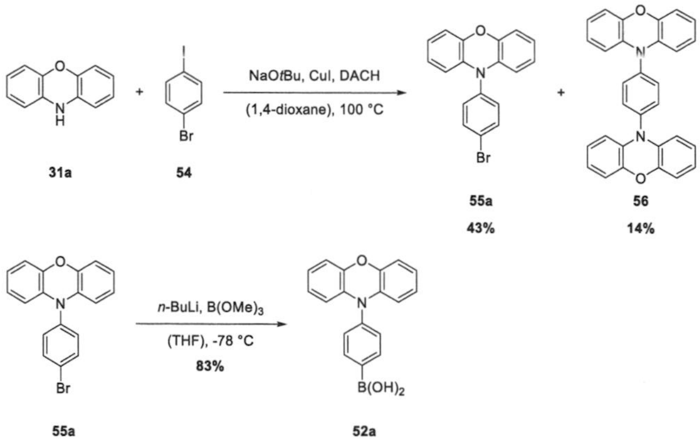

Scheme 17. Synthesis of precursor **52a**.

The synthesis of the DMAC-based precursor **52b** was conducted in an analogous manner as depicted in Scheme 18 according to a literature-known procedure.[151] The product was obtained as a pale-yellow solid in a yield of 12% over two steps.

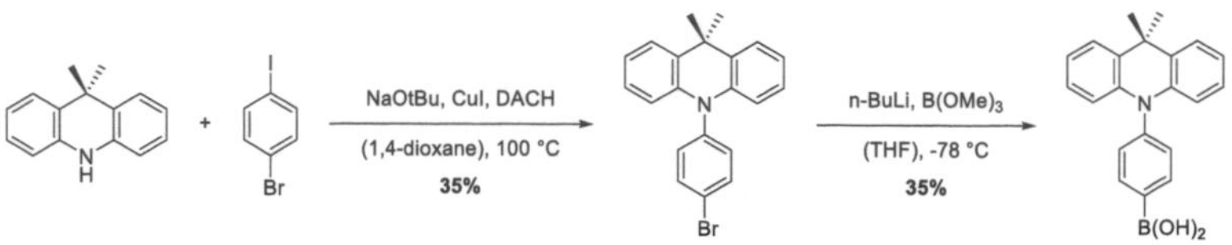

Scheme 18. Synthesis of precursor **52b**.

Following the precursor synthesis, SUZUKI coupling conditions were investigated. The results are summarized in Table 11, entries 1-7. The various reaction conditions employing different catalysts, ligands, temperatures, solvents, and reaction setups such as microwave irradiation vs. heating in a vial, were unsuccessful.

Table 11. Cross-coupling conditions for the synthesis of target compound **XIII** according to route B.

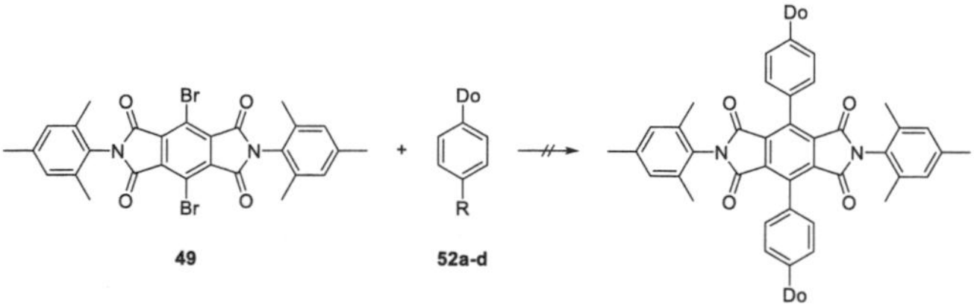

Entry	Reactant	R	Reaction Conditions
1	**52a**	B(OH)$_2$	Pd(PPh$_3$)$_4$, K$_2$CO$_3$, (H$_2$O/PhMe), 103 °C
2	**52a**	B(OH)$_2$	Pd(PPh$_3$)$_4$, K$_2$CO$_3$, (H$_2$O/THF), 79 °C
3	**52a**	B(OH)$_2$	PEPPSI-*i*Pr, KOtBu, (EtOH/PhMe), 100 °C
4	**52a**	B(OH)$_2$	Pd$_2$(dba)$_3$·CHCl$_3$, PCy$_3$, K$_3$PO$_4$, (1,4-dioxane), 100 °C
5	**52a**	B(OH)$_2$	Pd(OAc)$_2$, AntPhos, K$_3$PO$_4$, (1,4-dioxane), 140 °C
6	**52a**	B(OH)$_2$	Pd(PPh$_3$)$_4$, K$_2$CO$_3$, (1,4-dioxane/PhMe), 100 °C, MW, max. 250 W, 15 min
7	**52b**	B(OH)$_2$	Pd(PPh$_3$)$_4$, K$_2$CO$_3$, (H$_2$O/THF), 79 °C
8	**52c**	B(pin)	Pd(OAc)$_2$, AntPhos, K$_3$PO$_4$, (1,4-dioxane), 70 °C
9	**52d**	Sn(*n*-Bu)$_3$	Pd(PPh$_3$)$_4$, CuCl, LiCl, (DMF), 100 °C

Therefore, other coupling reactions such as a SUZUKI-MIYAURA and STILLE coupling were explored. Additional precursor synthesis of compounds **52c** and **52d** was conducted, as shown in Scheme 19, based on a stannylation reaction reported by Gribanov *et al.*[152] and according to a patent procedure by Zhou *et al.*[153]

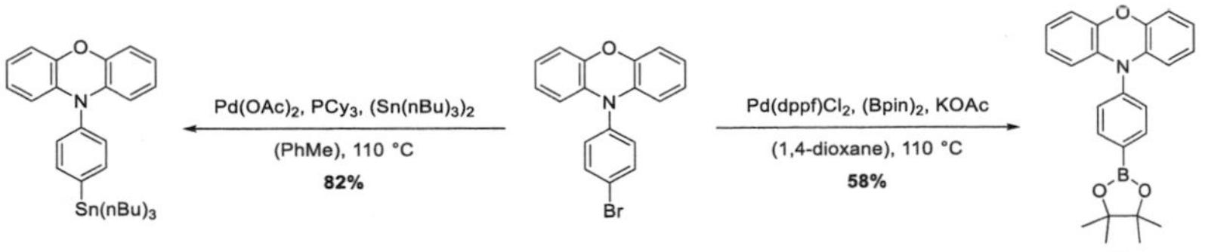

Scheme 19. Synthesis of precursors **52c** and **52d**.

The borate ester **52c** was obtained as a pale-yellow solid with a yield of 58%, while the tributylstannyl compound **52d** was synthesized as a colorless oil with a yield of 82%. After completing of the precursor synthesis, SUZUKI-MIYAURA and STILLE coupling were conducted with the reaction conditions displayed in Table 11, entries 8 and 9. However, both reaction pathways failed to synthesize compound **XIII** successfully. Both route A and B were ineffective in synthesizing target compound **XIII**. Therefore, the molecule design was re-adjusted.

In the initial publication that had inspired the PIT acceptor design, Adachi and coworkers had reported the linkage of the anthracene core to a benzene ring *via* an alkyne bond.[149] Thus, the synthesis of **PXZ-π-#-PIT-#-π-PXZ (57)** was investigated. Precursor **52f** was synthesized as a colorless oil *via* a SONOGASHIRA cross-coupling reaction and subsequent deprotection in a yield of 26% over two steps as depicted in Scheme 20.

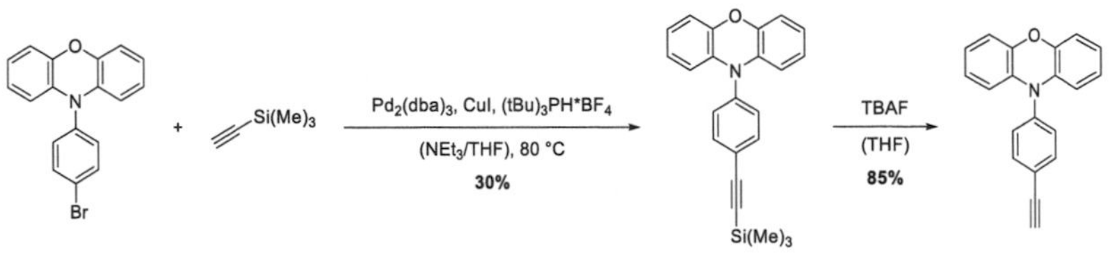

Scheme 20. Synthesis of precursor **52f**.

Following up on the successful synthesis of **52f**, a second SONOGASHIRA coupling reaction was conducted, as shown in Scheme 21. Target compound (**57**) was obtained as a dark-red solid with a yield of 78%.

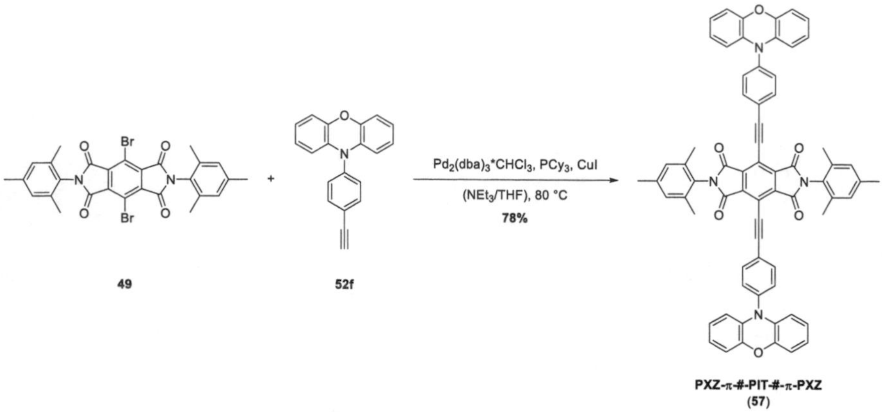

Scheme 21. Synthesis of **PXZ-π-#-PIT-#-π-PXZ** (**57**).

PIT – Absorption and Emission

In preliminary tests, the compound showed no luminescence when irradiated with a UV lamp at a wavelength of 365 nm at room temperature. However, when the sample was dissolved in toluene and cooled to 77 K, it showed a faint red luminescence under irradiation at 365 nm. UV-vis spectroscopy and fluorescence spectroscopy measurements were carried out to quantify the photophysical properties. The results thereof are displayed in Figure 32.

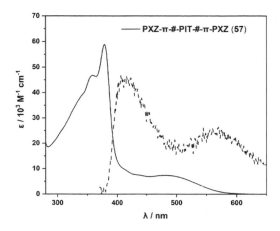

Figure 32. Absorption and normalized emission spectrum of **PXZ-π-#-PIT-#-π-PXZ** (**57**) in toluene (λ_{exc} = 350 nm).

The absorption measurement by means of UV-vis spectroscopy for **PXZ-π-#-PIT-#-π-PXZ** (**57**) shows three distinct peaks which were observed at 358, 378, and 488 nm in toluene solution under ambient conditions. The absorption spectrum is similar to the literature-reported spectra of its analogs, as reported by Kato et al.[154] and Kuila et al.[155] and the comparison of the spectral data of the three compounds (**57**), **59**, and **60** are summarized in Table 12.

Kuila et al.[155] reported a locally excited absorption band below 400 nm and another charge-transfer absorption band around 390-480 nm. The data reported by Kato et al.[154] as well as the data shown in Figure 32 are consistent with these findings. In comparison to compound **59**, compound (**57**) shows very similar absorption bands for the locally excited absorption. However, the absorption band attributed to the charge-transfer is bathochromically shifted by 70 nm through the addition of the donor-π-spacer-conjugate.

Table 12. Comparison of absorption and emission data with literature-known analogs to **PXZ-π-#-PIT-#-π-PXZ** (**57**) measured in toluene solution.

Entry	Variable	Unit	(57)	59[154]	60[155]
1	c	[mM]	0.11	0.10	0.05
2	λ_{max}	[nm]	358, 378, 487	353 , 375, 414	300−390, 390−480
3	ε	[10^3 M^{-1} cm^{-1}]	46.8, 58.9, 7.31	47.3, 67.5, 7.50	n/a
4	λ_{em}	[nm]	414, 565	477	350−525, 525−750

Moreover, the extinction coefficients ε are in the same range in comparison to the literature-reported compound **59**. Regarding the emission wavelengths, compound **59** shows green fluorescence in the region between 479-540 nm, while compound **60** shows locally excited emission in the range of 350 to 525 nm and a broad charge-transfer-induced emission band with a maximum wavelength of approx. 630 nm. Compound (**57**) showed two emission bands with maximum wavelengths of λ_{max} = 414 nm and 565 nm.

PIT – DFT Calculation and Electrochemical Behavior

TD-DFT calculation results for the PIT-based compound (**57**) were provided by Dr. David Hall (Zysman-Colman Group, University of St. Andrews) and are displayed in Figure 33. The geometry optimized structure of (**57**) shows that the HOMO and LUMO are localized on the PXZ donors and PIT acceptor, respectively. However, the HOMO and LUMO are spatially segregated and therefore do not show an overlap. The LUMO energy of the PIT acceptor was calculated to amount to -2.97 eV, while the HOMO and HOMO-1 are located at -4.96 eV and -4.98 eV, respectively, in the gas phase using the Pbe0/6-31G(d,p) functional. The HOMO-LUMO gap, therefore, amounts to 1.99 eV. For TADF characteristics, the energy gap between the lowest-lying singlet and triplet state, ΔE_{ST}, is indicative as it allows for an efficient RISC process. The excited-state calculation of compound (**57**), which was carried out using the TAMM-DANCOFF approximation (TDA) in the gas phase using the Pbe0/6-31G(d,p) functional, displays a minuscule ΔE_{ST} value of 0.02 eV and a low oscillator strength of $f = 0.002$ for the S_1 state. The energy of the lowest-lying singlet state was computationally estimated to be at 1.66 eV, while the lowest-lying triplet state was calculated to be at 1.64 eV. Therefore, the upconversion of triplet excitons into the S_1 state should be feasible.

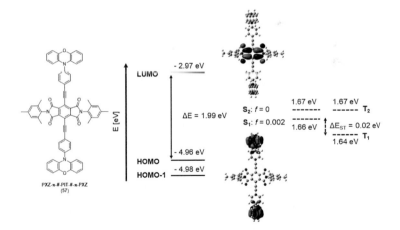

Figure 33. (TD)-DFT calculation results for (**57**) provided by Dr. David Hall (Zysman-Colman Group, University of St. Andrews); computation details: (i) geometry optimized structure by using the Pbe0/6-31G(d,p) functional, calculated in the gas phase, (ii) excited state calculated in the gas phase using the Pbe0/6-31G(d,p) functional, calculations performed using the TAMM-DANCOFF approximation (TDA).

In addition to the analysis by UV-vis and fluorescence spectroscopy and the computational efforts using TD-DFT methods, the electrochemical behavior was investigated by means of cyclic voltammetry to obtain experimental values for the HOMO and LUMO energies. Figure 34 shows the cyclic voltammogram with the reduction and oxidation processes which are both reversible. The oxidation and reduction potentials were 0.85 and -0.71 V vs. SCE, respectively. Thereby, the HOMO and LUMO levels were determined to be -5.19 and -3.63 eV, respectively, and the HOMO-LUMO gap amounts to 1.56 eV. The LUMO level is in the range of the literature-reported value for compound **59** by Kato *et al.* (E_{LUMO} = -3.132 eV). Besides the main reduction process, a second signal in the anodic current at -0.9 V vs. SCE is observed. It seems to be quasi-reversible with a matching cathodic current peak which is hinted at around 0.8 V vs. SCE. Interestingly, this signal is not observed after the addition of ferrocene as an internal standard.

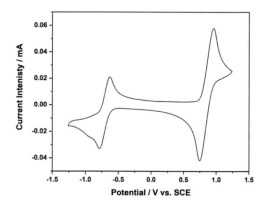

Figure 34. Cyclic voltammogram of compound (**57**) in Ar-saturated DCM solution (0.1 M [nBu$_4$N][PF$_6$]) at a scan rate of 100 mV s^{-1}.

The PIT-based acceptor system was established through continuous re-evaluation and adjustment of the molecule design. The synthesis of D-A-D-type, D-π-A-π-D-type using BUCHWALD-HARTWIG, SUZUKI-MIYAURA and STILLE cross-coupling were unsuccessful. However, a SONOGASHIRA cross-coupling could be realized, and the resulting compound (**57**) was characterized by means of UV-vis and fluorescence spectroscopy, TD-DFT computation and CV.

3.3. 1*H*-Pyrrolo[3,4-*b*]quinoxaline-1,3(2*H*)-dione (PQD)-Based TADF Molecule Design

PQD Derivatives and its Precursors – Synthesis

Following the work on the BP and PIT acceptor cores, a new acceptor design was established. The synthesis of molecules based on a 1*H*-pyrrolo[3,4-*b*]quinoxaline-1,3(2*H*)-dione (PQD) acceptor was initiated using a hypervalent I(III)-induced oxidative [4+2] annulation of phenylenediamine and electron-deficient alkyne as reported by Okumura *et al.* (Scheme 22).[156] The intermediate **62a** and **62b** were obtained in 56% and 17% yields, not reproducing the literature-reported yields of 85% and 89%, respectively. The procedure had been applied as reported with the difference that 3.00 equiv. of the reagent PhI(OAc)$_2$ was used instead of 2.00 equiv. Furthermore, the batch sizes were upscaled, which could explain the lower yields. In the case of **62a**, the batch size was upscaled to obtain the desired product on a gram scale, while the literature procedure reported a batch size below 100 mg. On the other hand, the batch size of **62b** was three times the reported batch size. Next, the synthesis of the precursor **63** was successfully conducted with an overall yield over two steps of 53%. In the case of intermediate **62b**, the nitrile group was expected to hydrolyze under the given conditions, and thus the synthesis pathway using this substrate was not further investigated based on a conflicting protecting group strategy.

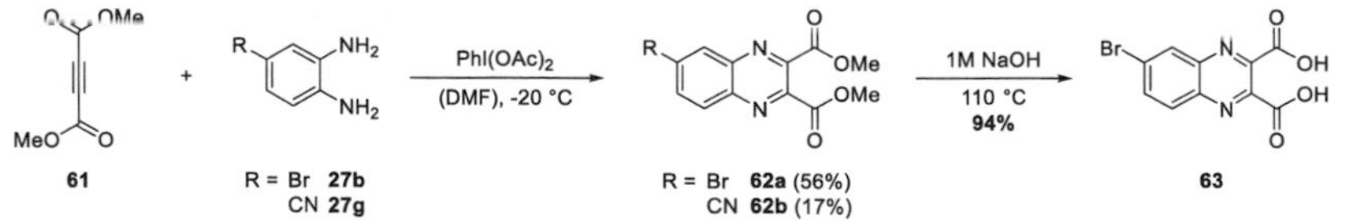

Scheme 22. Syntheses of PQD precursors **62a-b** according to Okumura *et al.*[156] and hydrolysis of **62a** to obtain **63**.

Subsequently, the reaction conditions leading to imide formation were optimized, as displayed in Table 13. First, the same reaction conditions as for the imide formation of the PIT acceptor were applied. However, refluxing the reaction mixture at 120 °C in acetic acid (entry 1), did not yield the desired product but resulted in the formation of side product **65** with 3% yield. This hinted at an insufficient activation of the carboxylic acid through anhydride formation to form the second C-N bond. Therefore, the activation of the carboxylic acid *via* an acyl chloride was attempted unsuccessfully (entry 2). Then, the activation by peptide coupling reagents such as EDC·HCl was investigated. As this did not lead to the desired product formation, HOBt was added as an additive for the active ester formation (entry 4). This led to the formation of the desired imide **64a** in a yield of 10% as a pale yellow-orange solid. To increase the yield, other peptide reagents were tested as well. While the HATU-mediated coupling led to **Br-PQD-Mes (64a)** in a yield of 65%, the coupling employing HBTU did not result in product formation.

Table 13. Optimization of the imide synthesis reaction conditions.

Entry	Reaction Conditions	Yield (64a) [%]	Yield 65 [%]
1	(AcOH), 120 °C	-	3
2	(COCl)$_2$, pyridine, (THF), 70 °C	-	-
3	EDC·HCl, DMAP, (DCM), r.t.	-	-
4	EDC·HCl, HOBt, DIPEA, (DCM), r.t.	10	-
5	HATU, DIPEA, (DMF), r.t.	65	-
6	HBTU, DIPEA, (DMF), r.t.	-	-

As HATU proved to be the most promising coupling reagent, the substrate scope of the imide formation and the condition given in Table 13, entry 5, was broadened, as shown in Scheme 23. Compounds **(64b-g)** were obtained in moderate to excellent yields. It was observed that the electron-donating strength and number of substituents on the diamine compounds **48b-g** increased the yield. **Br-PQD-TMS (64g)** had initially been planned as a precursor to compound **Br-PQD-C≡CH (64f)**, but as the yield was low, the synthesis of **(64f)** was attempted directly, and the yield of this reaction was significantly higher.

While the synthesis of **Br-PQD-OMe (64b)** had been realized *via* a different synthetic route by Hanaineh-Abdelnour and Salameh in 1999,[157] a structure search on SciFinder^n (as of 04/12/2022) did not yield in any results for compound **(64c)**, **(64f)**, **(64g)**, and **(64e)**; although for **Br-PQD-F (64e)**, an analogous structure featuring a chloride instead of the bromine substituent had been reported in 1972 by Augustin *et al.*[158] Thus, the synthesis of aromatic imides *via* the HATU-mediated coupling of dicarboxylated quinoxaline derivatives and aniline derivatives was successfully established in a facile, modular synthesis approach. This approach was applied to synthesize the donor-acceptor conjugate **Br-PQD-TPA (64d)**, which was obtained in a yield of 87% as an orange solid. To the best of my knowledge, compound (**64d**) is not literature-reported.

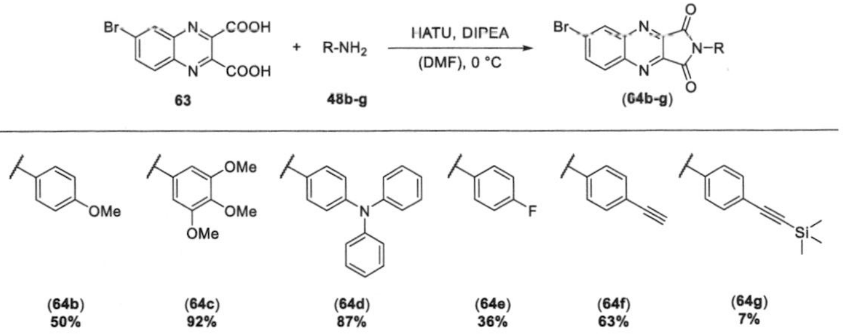

Scheme 23. Synthesis of PQD derivatives (**64b-g**).

In summary, the modular synthesis of aromatic imide-type PQD acceptors was successfully established by optimizating the reaction conditions. A first insight into the scope of this reaction was demonstrated using six substrates.

PQD Derivatives and its Precursors – Electrochemical Behavior

Subsequently, the compounds (**64a-g**) were investigated by means of cyclic voltammetry in order to determine experimental values for the LUMO levels (Figure 35) and thereby estimate the acceptor strength. The electrochemical measurements were conducted in Ar-saturated dichloromethane containing 0.1 M [nBu$_4$N][PF$_6$]. While the mesitylated (**64a**), *para*-fluorinated (**64e**), and 3,4,5-methoxy-bearing (**64c**) derivatives show reversible reduction processes, the alkyne-substituted PQD derivative (**64f**) shows a quasi-reversible reduction. In contrast, the *para*-methoxy-substituted derivative (**64b**) shows an irreversible reduction process.

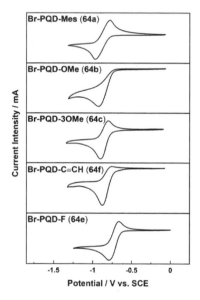

Figure 35. Cyclic voltammograms of the Br-PQD-R series (**64a-c**) and (**64e-f**) measured in Ar-saturated DCM containing 0.1 M [nBu$_4$F][PF$_6$] at a scan rate of 100 mV s^{-1}.

The reduction potentials and LUMO energies are summarized in Table 14. The lowest-lying LUMO is observed for the *para*-fluorine substituted PQD derivative (**64e**) at -3.61 eV. This is in accordance with literature-known design strategies that use fluorine substituents as auxiliary acceptors.[41] The alkyne-bearing compound (**64f**) also shows a stabilizing effect leading to a lowered LUMO level with an E_{LUMO} = -3.52 eV (Table 14, entry 4). The methyl and tri-methoxy substituted derivatives (**64a**) and (**64c**) display LUMO energies of -3.46 eV and -3.49 eV, respectively. The highest LUMO is found for **Br-PQD-OMe (64b)**, which shows an E_{LUMO} = -3.36 eV. The electron-accepting strength of the PQD acceptor core can hence be effectively tuned through substitution patterns and is individually adjustable in regards to the intended application.

Table 14. Reduction potentials and LUMO energies of the Br-PQD-R series (**64a-c**) and (**64e-f**).

Entry	Compound	E_{red} [V vs. SCE]	LUMO [eV]
1	**Br-PQD-OMe (64b)**	-0.98	-3.36
2	**Br-PQD-Mes (64a)**	-0.88	-3.46
3	**Br-PQD-3OMe (64c)**	-0.85	-3.49
4	**Br-PQD-C≡CH (64f)**	-0.82	-3.52
5	**Br-PQD-F (64e)**	-0.73	-3.61

The PQD-based acceptors (**64a-c**) and (**64e-f**) were characterized through cyclic voltammetry. The resulting data was used to experimentally estimate the LUMO energy levels and gain information regarding the electrochemical stability of the PQD derivatives. It was observed that **Br-PQD-F (64e)** showed the most stabilized LUMO level, and **Br-PQD-OMe (64b)** the least stabilized LUMO level. Out of the five investigated derivatives, three showed reversible reduction processes, one exhibited semi-reversibility, and one derivative displayed an irreversible reduction process.

PQD Derivatives and its Precursors - Absorption and Emission

The UV-vis measurement results of compounds (62a-b), (64a-c), and (64e-f) are visualized in Figure 36. While the precursors **Br-PQD-CO₂Me** (62a) and **CN-PQD-CO₂Me** (62b) (Figure 36a) show an excitation maximum at 337 and 400 nm, as well as 326, 338, and 431 nm, respectively, the PQD derivatives (Figure 36b) show maxima at longer wavelengths as summarized in Table 15. Both precursors show similar absorption bands between 320 and 340 nm. However, the absorption band at longer wavelengths is red-shifted for **CN-PQD-CO₂Me** (62b) by 31 nm compared to **Br-PQD-CO₂Me** (62a). This is due to the strong electron-withdrawing nature of the nitrile group, which stabilizes.

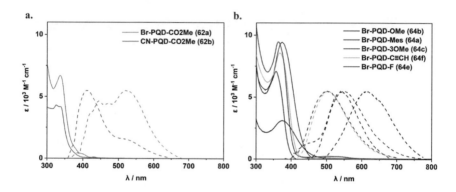

Figure 36. UV-vis and normalized fluorescence spectroscopy data of PQD precursors (62a-b) and PQD acceptors (64a-c, e-f) in toluene.

The absorption maxima of the imide derivatives are bathochromically shifted in dependence of the *N*-aryl substituent(s) in the following order: 2,4,6-Me < F < C≡C < OMe < 3,4,5-OMe. Moreover, the *para*-fluorinated compound (64e) shows additional absorption peaks at 506 and 535 nm, albeit with very low extinction coefficients of 0.23 and $0.22 \cdot 10^3 \ M^{-1} \ cm^{-1}$, respectively. The extinction coefficients for the Br-PQD-R series (Table 15) lie in a similar range except for **Br-PQD-3OMe** (64c), which exhibits a lower extinction coefficient than its peers.

The emission spectra of 0.1 mM solutions in toluene are indicated in Figure 36 in dotted lines. The precursor **Br-PQD-CO₂Me** (62a) was excited at a wavelength of $\lambda_{exc} = 350$ nm and shows three emission peaks. Two emission peaks are less defined and found at 443 nm and 471 nm, while the emission signal at 522 nm shows a Gaussian-like distribution. Furthermore, the precursor **CN-PQD-CO₂Me (62b)** showed two emission peaks at 411 and 519 nm. However, as previously mentioned, this precursor was not used in the following steps due to a conflict with the protecting group strategy.

Table 15. Summary of the data for the Br-PQD-R series obtained from UV-vis and fluorescence spectroscopy measurements in 0.1 mM toluene solution.

Entry	Compound	λ_{max} (ε) [nm] ($[10^3$ M^{-1} cm$^{-1}]$)	λ_{em} (λ_{exc}) [nm]
1	**Br-PQD-Mes (64a)**	357 (7.09)	455, 539 (350)
2	**Br-PQD-F (64e)**	362 (9.47), 506 (0.23), 535 (0.22)	502 (363)
3	**Br-PQD-C≡CH (64f)**	367 (8.58)	503 (367)
4	**Br-PQD-OMe (64b)**	374 (9.45)	547 (425)
5	**Br-PQD-3OMe (64c)**	376 (3.12)	611 (400)

The mesityl-substituted PQD derivative (**64a**) shows a weaker emission peak at 455 nm and a dominant emission signal at 539 nm when excited at λ_{exc} = 350 nm. In comparison to its precursor **Br-PQD-CO₂Me (62a)**, the emission of compound (**64a**) is red-shifted by 17 nm. Upon the introduction of a fluorine substituent to the system in **Br-PQD-F (64e)** (Table 15, entry 2), an emission is observed at 502 nm when excited with λ_{exc} = 363 nm. However, a different, bathochromically shifted emission wavelength of λ_{em} = 561 nm was detected when the sample was excited at λ_{exc} = 500 nm (chapter 8.1, Figure 60). Following compound (**64e**), Table 15, entry 3 shows the data for **Br-PQD-C≡CH (64f)**. The alkyne substituent on the compound (**64f**) led to an emission wavelength of 503 nm when excited at the absorption maximum. The *para*-methoxy substituted derivative **Br-PQD-OMe (64b)** red-shifts the emission wavelength further and displayed an emission maximum at 547 nm when excited at 425 nm. Lastly, the 3,4,5-methoxy-substituted derivative **Br-PQD-3OMe (64c)** displayed the most red-shifted emission with a maximum λ_{em} = 611 nm when excited at 400 nm.

In summary, as a general trend, with the exception of the emission of the fluorinated compound (**64e**), the absorption and emission maxima were bathochromically shifted in dependence on the number of substituents and the electron density they contribute. **Br-PQD-F (64e)** showed two emission wavelengths that could be selectively addressed by adjusting the excitation wavelength.

PQD-Based Donor and Acceptor Conjugates – Synthesis

After establishing the synthesis of the PQD acceptor derivatives and the D-A conjugate **Br-PQD-TPA** (**64d**) (Scheme 23), other D-A conjugates were synthesized. In contrast to **Br-PQD-TPA** (**64d**), the triphenylamine donor was connected to the PQD acceptor in a SUZUKI-MIYAURA cross-coupling reaction as shown in Scheme 24 to obtain **TPA-PQD-Mes** (**66a**) and **TPA-PQD-OMe** (**66b**) in a yield of 86 and 56% as a red solid and a yellow-orange solid, respectively. Both compounds show luminescence upon irradiation at 365 nm at room temperature.

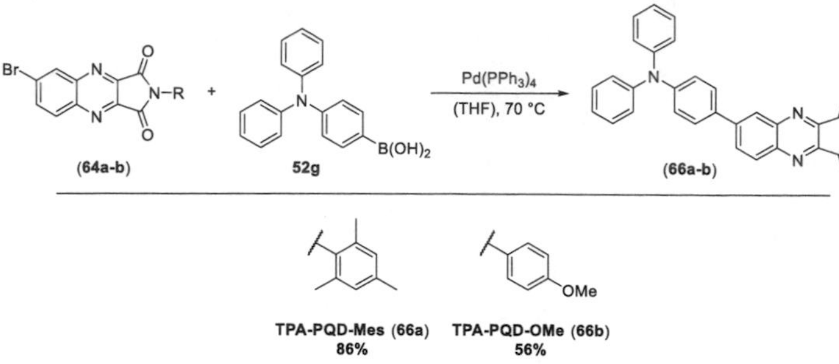

Scheme 24. Synthesis of donor-acceptor conjugates (**66a**) and (**66b**) based on the PQD acceptor.

In addition to the TPA donor, PXZ **31a** was also investigated for PQD-based D-A systems. As shown in Scheme 25, a SUZUKI-MIYAURA cross-coupling was carried out to obtain **PXZ-PQD-Mes** (**67**) as a dark violet-black solid with a yield of 62%. When dissolved in ethyl acetate, the solution was red. However, any attempts to produce luminescence at room temperature under irradiation with λ = 365 nm were unsuccessful. Therefore, the sample was dissolved in toluene, cooled to 77 K, and then irradiated with the abovementioned wavelength. Bright yellow luminescence was observed.

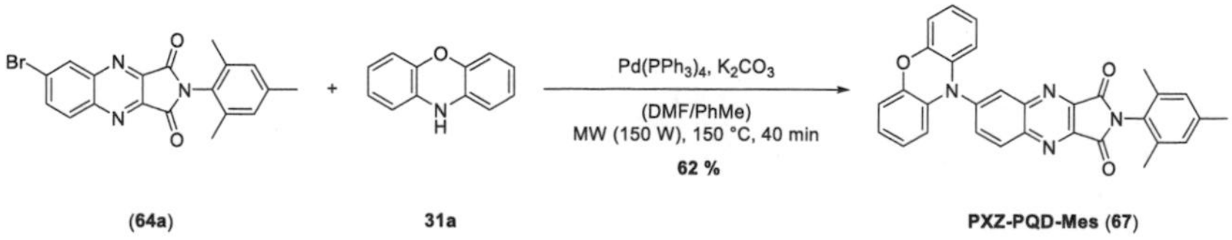

Scheme 25. Synthesis of D-A-type emitter **PXZ-PQD-Mes**.

The molecule design was extended by adding a π-spacer to investigate the effect of more separated frontier molecular orbitals. It had been reported that an increased separation of the donor and acceptor led to an increased radiative rate constant, k_r.[159] The synthesis of **PXZ-π-BP-Mes (68)** was realized through a SUZUKI-MIYAURA cross-coupling in 68% as a yellow solid, as depicted in Scheme 26. The purification *via* column chromatography was tedious and the solubility in a cyclohexane/ethyl acetate solvent mixture was an issue. No fluorescence could be observed when irradiated with light of λ = 365 nm at room temperature. Interestingly, at 77 K, yellow-orange luminescence was observed.

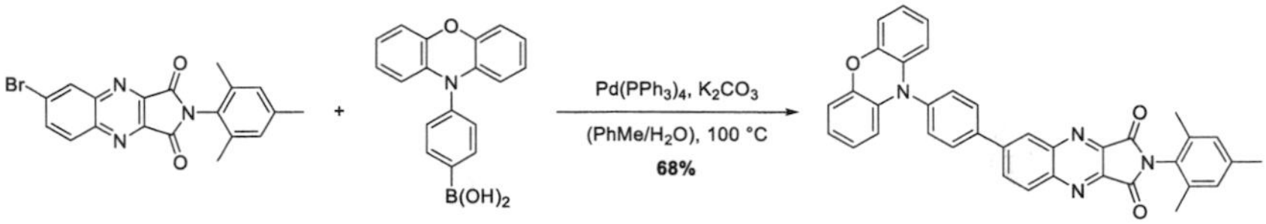

Scheme 26. Synthesis of D-π-A-type structure **PXZ-π-PQD-Mes (68)**.

The NMR analysis of **PXZ-π-PQD-Mes (68)** showed rotamers. Figure 37 shows the ^{1}H NMR spectrum and a close-up of the aromatic region between 5.50–9.00 ppm. The ^{1}H NMR spectrum shows impurities. Solvent signals of dichloromethane and ethyl acetate are observed at δ = 5.30 ppm and δ = 4.12 and 2.05 ppm, respectively. Furthermore, grease is observed at δ = 1.27 ppm and 0.86 ppm.[160]

First, the more prominent signals belonging to the main rotamer, indicated with the index "a", were identified. The protons 2-H, 7-H and 8-H were assigned based on their integrals, ^3J, and ^4J coupling constants (see chapter 5.2.3), and the cross-peaks found in the COSY NMR spectrum shown in Figure 38a. The HSQC NMR spectrum assigns C-2, C-7 and C-8 to the ^{13}C signals at δ = 128.63, 132.00 and 133.16 ppm, respectively. The signals of C-3 and C-6 were identified *via* HMBC NMR and determined to δ = 145.10 and 144.10 ppm, respectively. Furthermore, the remaining signals of C-4, C-5, and the carbonyl signals are stated in chapter 5.2.3. The signals of the mesityl group are also listed in chapter 5.2.3. However, for the discussion of the ^{1}H NMR spectrum and the occurrence of rotamers, it is of interest to mention that the singlet of 3'-H was observed at 7.07 ppm.

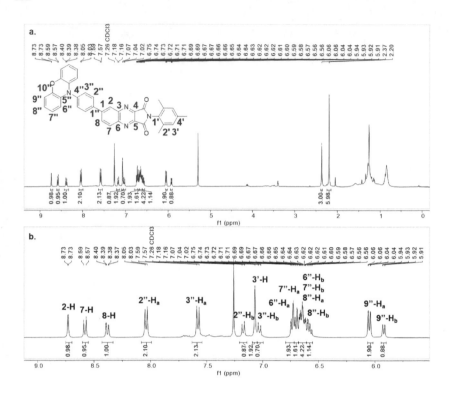

Figure 37. ^{1}H NMR of **PXZ-π-PQD-Mes** (**68**) in CDCl$_3$, **b.** excerpt of **a.** between 5.1-9.2 ppm.

After the PQD core and the mesityl substituent were characterized by NMR, the protons belonging to the π-spacer were assigned. 2"-H$_a$ and 3"-H$_a$ were identified through COSY NMR. 3"-H$_a$ was assigned through a ^3J cross-peak with C-1 in the HMBC NMR spectrum (Figure 38b). The ^{13}C NMR signal of C-1 was cross-validated through 7-H.

Therefore, at this point, the PXZ-protons were left to be assigned. Due to the proximity to the oxygen atom, which shows a higher electronegativity than nitrogen, 9"-H is anticipated to be shielded and show an upfield shift compared to 6"-H. Thus, with regards to the integral, 9"-H$_a$ was assigned to δ = 6.05 ppm. The COSY NMR spectrum, displayed in Figure 38c, located the 8"-H$_a$ via ^3J coupling with 9"-H$_a$ in the multiplet between δ = 6.75–6.54 ppm. Furthermore, it showed that the multiplet in this range displays at least four signals out of which the signal for 8"-H$_a$ was more specified to be found in the multiplet between δ = 6.68–6.61 ppm. Next, the signal for the remaining 7"-H$_a$ and 6"-H$_a$ were investigated. The signal of 7"-H$_a$ was assigned employing HMBC and HSQC NMR. The HMBC NMR spectrum shows a cross-peak of with 9"-H$_a$ two carbon signals at δ = 144.14 ppm and 121.85 ppm. The former belongs to C$_a$-5" as it does not show a peak in the HSQC NMR spectrum, while the latter can thus be assigned to C$_a$-7" through which the signal of 7"-H$_a$ is traced back in the HSQC NMR spectrum to be found as a doublet of doublets at δ = 6.70 ppm which also cross-couples with C$_a$-5" in the HMBC NMR spectrum (Figure 38d). Hence, only 6"-H$_a$ was left to be identified in the ^{1}H NMR spectrum. Both 6"-H$_a$ and 8"-H$_a$, which are located in the signal pool between δ = 6.75–6.54 ppm, should be showing cross-peaks with C$_a$-10" in the HMBC NMR. In Figure 38d (red line), two such occurrences are found. Both feature the proton represented by the doublet of doublets at δ = 6.74 ppm and the multiplet where 8"-H$_a$ is located. Thus, the former signal was identified as 6"-H$_a$. Through these findings, C$_a$-10" was assigned to the peak at δ = 134.16 ppm as it does not show a signal in the HSQC NMR.

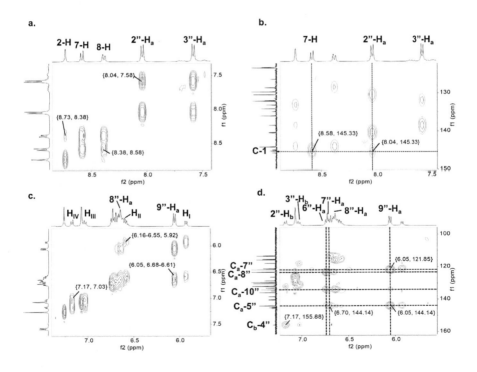

Figure 38. a. and **c.** COSY NMR, and **b.** and **d.** HMBC NMR spectra of **PXZ-π-PQD-Mes (68)** in CDCl$_3$.

After assigning the signals belonging to **PXZ-π-PQD-Mes** (**68**), the unidentified signals were investigated. The occurrence of a rotamer was considered besides the observation of unreacted reactant **PXZ-π-B(OH)₂** (**52a**) or defunctionalization thereof. However, the defunctionalization to **PXZ-π-H** could be ruled out based on the absence of the proton, which would have replaced the boronic acid functional group, and was reported to appear as a triplet at δ = 7.48 ppm measured under the same conditions in CDCl₃ measured at 400 MHz.[161] Furthermore, the presence of unreacted precursor (**52a**) is highly unlikely after two rounds of purification by chromatography, and was ruled out based on reported NMR data.[150]

Therefore, the possibility of a rotamer was explored. The COSY NMR spectrum displayed in Figure 38c shows ^3J coupling between H$_I$ and H$_{II}$, as well as H$_{III}$ and H$_{IV}$. H$_{III}$ and H$_{IV}$ are doublets at δ = 7.17 and 7.03 ppm and could be assigned to 2"-H$_b$ and 3"-H$_b$, respectively. HSQC NMR data (Figure 40a) assigned the C$_b$-2"and C$_b$-3" to δ = 132.11 ppm and 117.92 ppm, respectively. The HMBC NMR spectrum subsequently showed ^3J cross-coupling to C$_b$-4" at δ = 155.88 ppm and C$_b$-1" at δ = 131.50 ppm, respectively (Figure 38d). A less intense ^2J coupling of C$_b$-4" and 3"-H$_b$ was also observed. Unfortunately, the cross-peak between 2"-H$_b$ and C-1 was not observed.

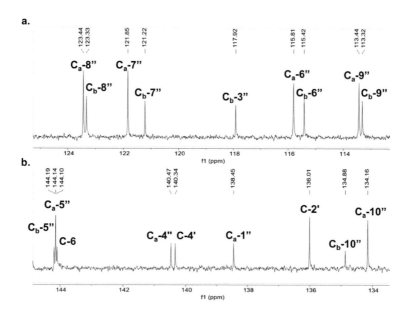

Figure 39. ^{13}C NMR extracts of **PXZ-π-PQD-Mes** (**68**) in CDCl₃.

In addition to this set of ^{1}H and ^{13}C NMR signals, the resemblance of H_I and 9"-H_a and the upfield shift indicated an additional set of 10H-phenoxazine protons, whereas H_I could be assigned to 9"-H_b. The COSY NMR (Figure 38c) showed ^3J coupling to the multiplet between δ = 6.61–6.55 ppm, which was thereby assigned to 8" H_b. The multiplet between δ = 6.68–6.61 ppm was thus far associated with 8"-H_a but showed an integral > 4. Thus, by comparison with the COSY NMR data, the remaining protons 6"-H_b and 7"-H_b were allocated to this multiplet. Furthermore, the rotamer signals can be observed through the ^{13}C NMR data (Figure 39). HSQC NMR assigned the C_b-9", C_b-8", C_b-7", and C_b-6" to δ = 113.32, 123.33, 121.22, and 115.42 ppm, respectively (Figure 40a). Lastly, C_b-5" and C_b-10" were assigned to δ = 144.19 ppm and 134.88 ppm through HMBC NMR (Figure 40b) analysis and cross-peaks with 9"-H_b and 8"-H_b, respectively.

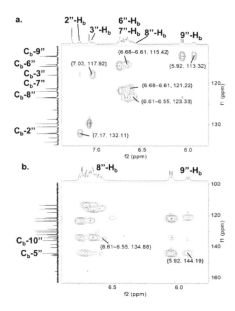

Figure 40. a. HSQC NMR spectrum and **b.** HMBC NMR spectrum of **PXZ-π-PQD-Mes (68)**.

Four donor- and acceptor conjugates were synthesized based on the PQD acceptor. In contrast to the D-A-type conjugates **TPA-PQD-Mes (66a)**, **TPA-PQD-OMe (66b)** and **PXZ-PQD-Mes (67)**, D-π-A-type **PXZ-π-PQD-Mes (68)** was obtained as two rotamers, which 1D and 2D NMR analysis confirmed. Similar acceptor cores have recently been published by Bin and coworkers and indicate the high potential of new aromatic imide-based acceptors for TADF applications.[46]

113

PQD-Based Donor and Acceptor Conjugates – DFT Calculations and Electrochemical Behavior

For the PQD-based emitter design, TD-DFT calculations were conducted by Changfeng Si (University of St. Andrews, Zysman-Colman group). The results for the molecules **PXZ-π-PQD-Mes (68)** and **TPA-PQD-Mes (66a)**, and **TPA-PQD-OMe (66b)** are shown in Figure 41. The ground-state geometry optimization shows that the HOMO is mainly located on the 10*H*-phenoxazine and triphenylamine donors, while the LUMO is mainly located on the PQD acceptor and spreads out slightly onto the neighboring phenyl spacer. As expected, the stronger donor PXZ shows a shallower HOMO level (E_{HOMO}(PXZ) = -5.08 eV) than TPA (E_{HOMO}(TPA) = -5.44 eV). Interestingly, the PQD acceptor for the D-π-A system depicts a lower LUMO (-2.78 eV) than in the D-A system (-2.55 eV). A more efficient extended conjugation onto the phenyl spacer might be the reason for stabilizing the LUMO. As a result, **PXZ-π-PQD-Mes (68)** and **TPA-PQD-Mes (66a)** show a HOMO-LUMO gap of ΔE = 2.24 eV and ΔE = 2.90 eV, respectively. However, a stark difference is visible in the excited state as **PXZ-π-PQD-Mes (68)** displays a minuscule ΔE_{ST} = 0.01 eV in combination with an oscillator strength of f = 0.0134, while **TPA-PQD-Mes (66a)** exhibits a ΔE_{ST} = 0.30 eV and an oscillator strength of f = 0.28. These results indicate that **TPA-PQD-Mes (66a)** is likely to show less efficient reverse intersystem crossing and TADF properties.

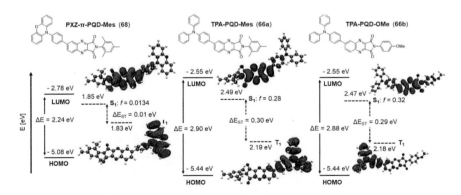

Figure 41. Geometry optimization and excited state (TD)-DFT calculation results for **PXZ-π-PQD-Mes (68)**, **TPA-PQD-Mes (66a)** and **TPA-PQD-OMe (66b)** provided by Changfeng Si (Zysman-Colman Group, University of St. Andrews); computation in the gas phase using the Pbe1pbe/6-31G(d,p) functional.

114

In comparison to **TPA-PQD-Mes (66a)**, **TPA-PQD-OMe (66b)** only varies in the excited state computation, where it shows slightly different S_1 and T_1 levels, as well as a higher oscillator strength of $f = 0.32$, and a $\Delta E_{ST} = 0.29$ eV. However, it does not show highly divergent computational results compared to **TPA-PQD-Mes (66a)**, whereby it can be deduced that the methoxy substituent does not heavily influence the electronic nature of the TPQ-PQD-conjugate.

To obtain experimental values for the HOMO and LUMO of the TPA-PQD-R series, cyclic voltammetry was conducted. The results thereof are shown in Figure 42. In contrast to the DFT calculations which considered a molecule in the gas phase, the CV experiments were carried out in an electrochemical cell with electrodes and the electrolyte tetrabutylammonium hexafluorophosphate in Ar-saturated dichloromethane solution.

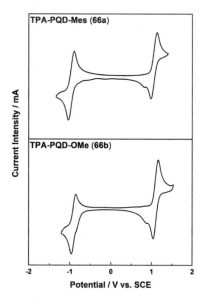

Figure 42. Cyclic voltammograms of the D-PQD-R-type emitters **TPA-PQD-Mes (66a)** and **TPA-PQD-OMe (66b)** in Ar-saturated DCM containing 0.1 M [nBu$_4$N][PF$_6$] at a scan rate of 100 mV s^{-1}.

Both **TPA-PQD-Mes (66a)** and **TPA-PQD-OMe (66b)** show reversible oxidation and reduction processes. The oxidation potentials were observed at E_{ox}(**66a**) = 1.06 V vs. SCE and E_{ox}(**66b**) = 1.10 V vs. SCE. Hence, the HOMO levels were calculated to be -5.40 and -5.44 eV for compound (**66a**) and (**66b**). This is in very good accordance with the values obtained from the DFT calculation. On the other hand, the reduction potentials were found to be at E_{red}(**66a**) = -0.95 V vs. SCE and E_{red}(**66b**) = -0.91 V vs. SCE. Thereby, the LUMO energies were -3.39 eV and -3.43 eV. These results show a more stabilized LUMO than obtained in the DFT calculation. As the DFT calculations were carried out in the gas phase, this might be attributed to a solvent effect in dichloromethane. Consequently, the experimental value for the HOMO-LUMO gap was determined as 2.01 eV for both compounds. The oxidation and reduction potentials of **TPA-PQD-Mes** (**66a**) and **TPA-PQD-OMe** (**66b**) and their experimental HOMO/LUMO values lie in a very similar range. They indicate that the substituent at the imide position does not immediately affect the electronic nature of the electron-accepting moiety.

In summary, TD-DFT calculation of the excited states shows that the likelihood of efficient thermally activated delayed fluorescence is highest for **PXZ-π-PQD-Mes** (**68**), in comparison to **TPA-PQD-Mes** (**66a**), **TPA-PQD-OMe** (**66b**). Furthermore, the experimentally obtained value for the HOMO indicates that the donor moiety is in accordance with the calculated value, while the experimental LUMO value is lower than its calculated counterpart.

PQD-Based Donor and Acceptor Conjugates – Absorption and Emission

In addition to the investigation employing TD-DFT calculation and CV, the absorption and emission processes occurring for the PQD-based donor and acceptor conjugates were investigated by means of UV-vis and fluorescence spectroscopy. Figure 43 shows the extinction coefficients against the wavelength, which indicate the absorption processes, while the dotted lines represent the results obtained from fluorescence spectroscopy. The emission spectra were obtained from 0.1 mM solutions in toluene.

The data obtained for the PXZ-based molecules, **PXZ-PQD-Mes (67)** and **PXZ-π-PQD-Mes (68)**, is shown in Figure 43a. Both compounds show very low extinction coefficients indicating inefficient absorption. The absorption bands below 400 nm show peaks at 319 and 373 nm for **PXZ-PQD-Mes (67)** and **PXZ-π-PQD-Mes (68)**, respectively. Moreover, **PXZ-PQD-Mes (67)** shows broad absorption bands at 461 and 543 nm, while **PXZ-π-PQD-Mes (68)** displays an absorption band at 483 nm. Furthermore, emission data employing fluorescence spectroscopy of **PXZ-PQD-Mes (67)** could not be obtained, while **PXZ-π-PQD-Mes (68)** displayed an emission wavelength maximum at 677 nm, albeit with very low intensity.

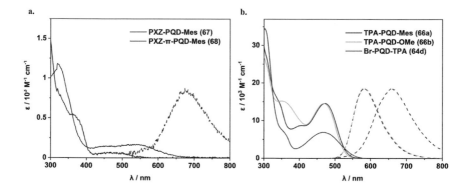

Figure 43. UV-vis spectra and normalized fluorescence spectra (shown in dotted lines) measured in toluene for the D-PQD-R-type **TPA-PQD-Mes (66a)** and **TPA-PQD-OMe (66b)**, as well as **Br-PQD-TPA (64d)**.

The absorption and emission spectra of the two TPA-PQD-based compounds **(66a-b)** and the D-A-type compound **(64d)** are depicted in Figure 43b. The data that was obtained from the UV-vis and fluorescence spectroscopy is summarized in Table 16. All three emitters show pronounced absorption bands above 410 nm which can be ascribed to charge-transfer processes from the donor to the PQD acceptor. When comparing the TPA-PQD-based emitters (66a) and (66b), which showed very similar behavior in the CV measurements, it shows that they have similar absorption bands around 285-290 nm and approximately 470 nm, but show different absorption maxima and extinction coefficients in the region between 340-410 nm. The mesitylated compound (66a) shows a bathochromically shifted absorption band with λ_{max} = 406 nm, in regards to compound (66b), which bears a methoxy substituent and shows a λ_{max} = 349 nm. The absorption maximum was shifted by 57 nm.

A difference can also be observed in the extinction coefficient values. While the absorption maxima above 410 nm show almost identical extinction coefficients of $14.54 \cdot 10^3$ M^{-1} cm^{-1} and $14.53 \cdot 10^3$ M^{-1} cm^{-1} for **TPA-PQD-Mes (66a)** and **TPA-PQD-OMe (66b)** (Table 16), the ε for the two absorption bands visible before 410 nm differ. The extinction coefficient of compound (66b) at a wavelength of 349 nm is $15.17 \cdot 10^3$ M^{-1} cm^{-1} and significantly higher than the extinction coefficient of compound (66a), which is $8.67 \cdot 10^3$ M^{-1} cm^{-1} at the absorption maximum at 406 nm. The absorption bands below 300 nm exhibit extinction coefficients in a similar range for all three TPA-based compounds (Table 16, entry 3-5).

Table 16. Summary of the data obtained for the PQD-based donor and acceptor conjugates from the UV-vis and fluorescence spectroscopy measurements in 0.1 mM toluene solutions.

Entry	Compound	λ_{max} (ε) [nm] ([10^3 M^{-1} cm^{-1}])	λ_{em} (λ_{exc}) [nm]
1	PXZ-PQD-Mes (67)	319 (1.18), 461 (0.14), 543 (0.17)	-
2	PXZ-π-PQD-Mes (68)	373 (0.51), 483 (0.07)	677 (500)
3	TPA-PQD-Mes (66a)	288 (34.15), 406 (8.67), 473 (14.54)	581 (465)
4	TPA-PQD-OMe (66b)	287 (31.61), 349 (15.17), 469 (14.53)	587 (469)
5	Br-PQD-TPA (64d)	303 (34.42), 360 (6.51), 466 (6.88)	662 (472)

The emission of the donor and substituent-bearing PQD derivatives was measured by means of fluorescence spectroscopy and the results are summarized in Table 16. **Br-PQD-TPA (64d)** showed the furthest red-shifted emission at 662 nm, while its D-A-type analogues **TPA-PQD-Mes (66a)** and **TPA-PQD-OMe (66b)** with a different donor-to-acceptor position displayed emission maxima at 581 nm and 587 nm, respectively. This is due to the different connectivity of the donor unit to the PQD acceptor for **Br-PQD-TPA (64d)**. For the TPA-

PQD-R-type molecules, the HOMO and LUMO overlap is larger. Thus, the acceptor is weakened as the HOMO feeds electron-density into the electron-accepting moiety. However, when connected *via* the imide position, the electronic coupling is significantly reduced and thus the FMOs are more effectively separated and the acceptor is thus stronger.

The PQD-based D-A-type and D-π-A-type compounds **PXZ-PQD-Mes** (**67**) and **PXZ-π-PQD-Mes** (**68**) did not show efficient absorption or emission behavior. Although the emission of the latter compound was heavily red-shifted into the near-infrared spectral region, the intensity of the emission was observed to be very low. This is likely due to non-radiative processes whose competitive impact gain weight as the energy difference between the lowest-lying excited singlet state and ground state becomes smaller. Furthermore, **TPA-PQD-Mes** (**66a**) and **TPA-PQD-OMe** (**66b**), and **Br-PQD-TPA** (**64d**).

3.4. Red-Shifted Azo-Menthol Derivatives

The following results were obtained during a research stay at the Trauner group at New York University (NYU) from September to December 2021. The focus of the contribution from my side regarding this project was the synthesis of red-shifted **Azo-Menthol** derivatives. The **Azo-Menthol (69)** structure design and synthesis had been previously established in the Trauner group by Dr. David Konrad (now at LMU Munich). This chapter is subdivided into three subchapters, highlighting the synthesis of Azo-Menthol derivatives with different connectivity between the menthyl and azobenzene moiety.

3.4.1. Azo-Menthyl-*C*-Amide Derivatives

Following the work of Dr. David Konrad, analogs to **Azo-Menthol (69)**, compounds **(70a-c)**, with an amide bond connecting the menthyl and azobenzene derivatives, were targeted and synthesized from the commercially available (−)-menthylformic acid **71** and azobenzenes **72a-c** as shown in Scheme 27. As the carbonyl C atom is attached to the menthyl moiety, this series of compounds was titled Azo-Menthyl-*C*-amide derivatives.

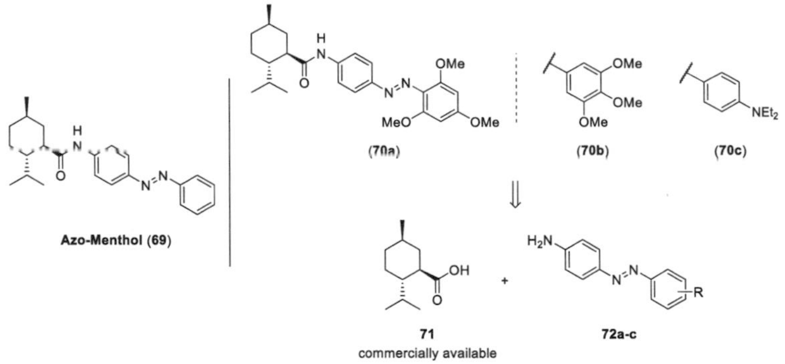

Scheme 27. The structure of Azo-Menthol **(69)** and the retrosynthetic outline of compounds **(70a-c)** based on (−)-menthylformic acid **71** and azobenzenes **72a-c**.

Azo-benzenes **72a-c** were synthesized starting from 4-nitroaniline **48h** *via* the formation of the diazonium compound, which was followed up by the respective azo coupling reactions. A subsequent reduction of the nitro group to the amine group resulted in the synthesis of compounds **72a-c**.

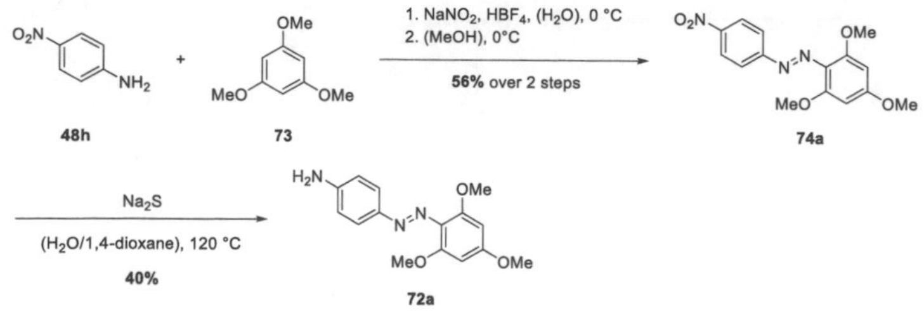

120

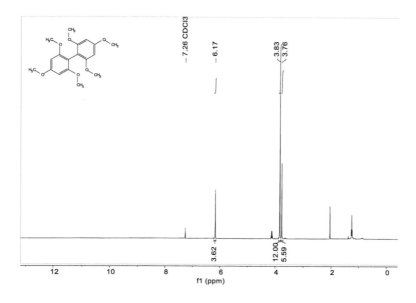

Scheme 28. Synthesis of azobenzene **72a**.

The synthesis of compound **74a** was achieved with a yield of 56% *via* a diazonium salt formation followed by an azo coupling reaction as depicted in Scheme 28. Subsequently, the reduction of the nitro group using SnCl$_2$ (4 eq.) in an HCl/methanol (1:1 vol%) mixture while stirring at 80 °C was unsuccessful. Instead of the reduction, after the purification of the crude material, 2,2',4,4',6,6'-hexamethoxy-1,1'-biphenyl **75** was observed in the ^{1}H NMR spectrum, as shown in Figure 44.

Figure 44. ^{1}H NMR of 2,2',4,4',6,6'-hexamethoxy-1,1'-biphenyl **75**.

Therefore, the reaction conditions were optimized, and disodium sulfide in water/1,4-dioxane (1:3) was used to successfully synthesize compound **72a** in a yield of 40% as shown in Scheme 28.

In addition to the 2,4,6-trimethoxy-substituted azobenzene **72a**, the 3,4,5-trimethoxy-substitued azobenzene **72b** was synthesized. The synthesis route to compound **74b** was analogous. However, **74b** showed difficulties regarding its solubility and was therefore reacted further without a purification step as depicted in Scheme 29. Then, iodomethane was used as a methylating agent to install the third methoxy substituent at the azobenzene before reducing the nitro functional group to the amine moiety. Compound **72b** was obtained in an overall yield of 58% over four steps starting from 4-nitroaniline **48h**.

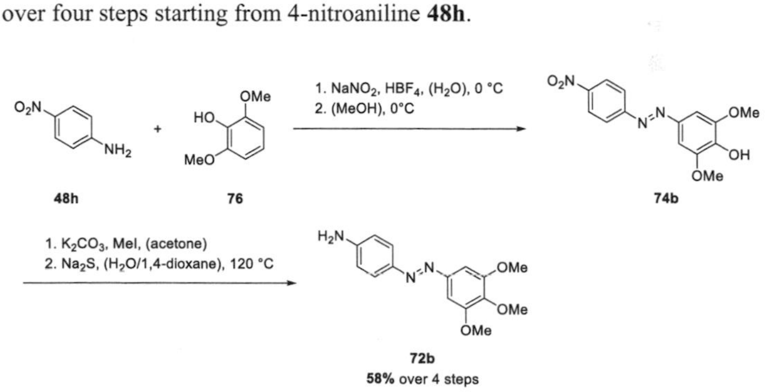

Scheme 29. Synthesis of azobenzene **72b**.

Furthermore, a third azobenzene bearing a diethylamine group, compound **72c**, was synthesized as shown in Scheme 30. Following a literature procedure by Trauner and co-workers[162], the intermediate **74c** was synthesized. The low yield of 7% was due to suboptimal reaction times in the first and the second step. The reaction leading to 1-nitro-4-nitrosobenzene was stirred overnight, while it was observed in previous experiments that for this step, with 4-nitroaniline **48h** as a starting material, three to five hours of reaction time seemed to be ideal. The subsequent BAEYER-MILLS reaction was conducted within 15 hours. Compound **72c** was successfully obtained after a reduction step with disodium sulfide in a yield of 61%.

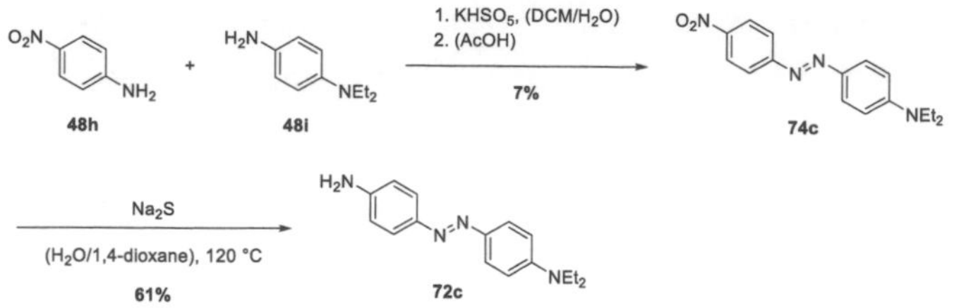

Scheme 30. Synthesis of diethyl amine-substituted azobenzene **72c**.

After the three azobenzenes **72a-c** were successfully synthesized, the respective amide coupling reactions with (−)-menthylformic acid **71** were conducted to obtain the azo-menthol derivatives (**70a-c**).

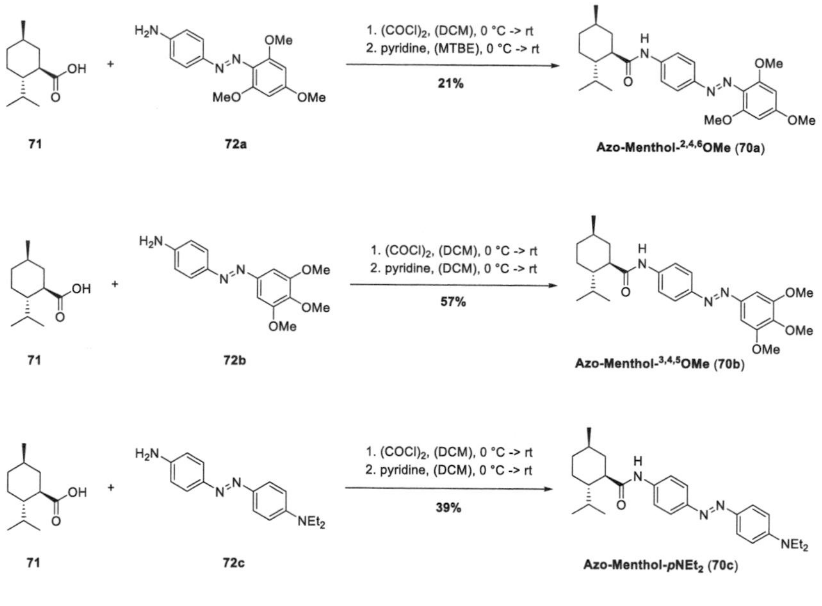

Scheme 31. Synthesis of azo-menthyl-C-amide derivatives (**70a-c**).

Scheme 31 summarizes the synthesis of **Azo-Menthol-2,4,6OMe (70a)**, **Azo-Menthol-3,4,5OMe (70b)**, and **Azo-Menthol-pNEt2 (70c)** starting from (−)-menthylformic acid **71** *via* an acyl chloride formation and consecutive amide coupling with azobenzenes **72a-c** under basic conditions. The yields of the reactions leading to the formation of **Azo-Menthol-2,4,6OMe (70a)**,

Azo-Menthol-3,4,5OMe (70b), and **Azo-Menthol-pNEt2 (70c)** were 21%, 57% and 39%, respectively. The stereochemistry for compound **(70b)** was confirmed using NOESY NMR (chapter 8.1, Figure 61).

3.4.2. Azo-Menthyl-N-Amide Derivatives

Inspired by the work by Journigan et al.[76] and Ortar et al.[77], who investigated TRPM8 antagonists such as **(−)-menthyl 1 (21)** (Scheme 32), the red-shifted azo-menthyl-N-amide derivative **77** was synthesized from (−)-methylamine **78** and the azobenzene **79b**.

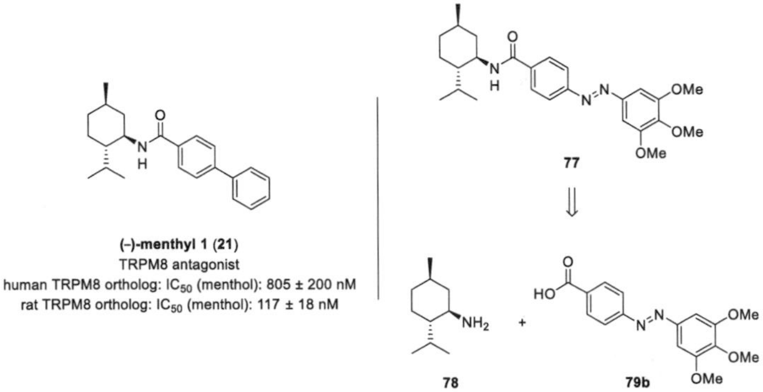

(-)-menthyl 1 (21)
TRPM8 antagonist
human TRPM8 ortholog: IC$_{50}$ (menthol): 805 ± 200 nM
rat TRPM8 ortholog: IC$_{50}$ (menthol): 117 ± 18 nM

Scheme 32. Structure of **(−)-menthyl 1 (21)** as reported by Journigan et al.[76] and the azo-menthyl-N-amide derivative target structure **(77)**.

The synthesis of (−)-menthylamine **78** was carried out according to a literature-known procedure from (L)-menthone via the oxime **81**, as shown in Scheme 33.[76] The first reaction step was carried out with a yield of 96%. The subsequent reduction reaction led to the desired product. However, as (−)-menthylamine **78** is volatile and unstable, it was used without further purification.

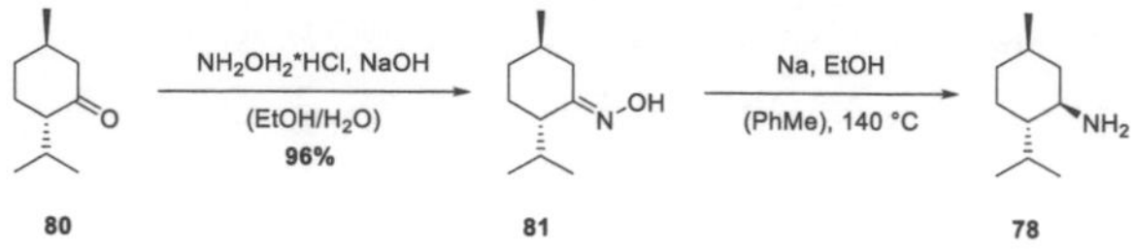

Scheme 33. Synthesis of (-)-menthylamine **78**.

For the synthesis of azo-menthyl-*N*-amides, azobenzenes bearing carboxylic acid functional groups **79a-c** were synthesized. As depicted in Scheme 34, the respective esters **82a** and **82b** were synthesized *via* azo coupling reactions in 53 and 68% yield, respectively. While compound **82a** was quantitatively hydrolyzed to give azobenzene **79a**, the hydroxyl group of compound **82b** was first methylated to give compound **82c** with a yield of 80%, before the carboxylic acid-substituted azobenzene **79b** was obtained through near quantitative hydrolysis of the methyl ester.

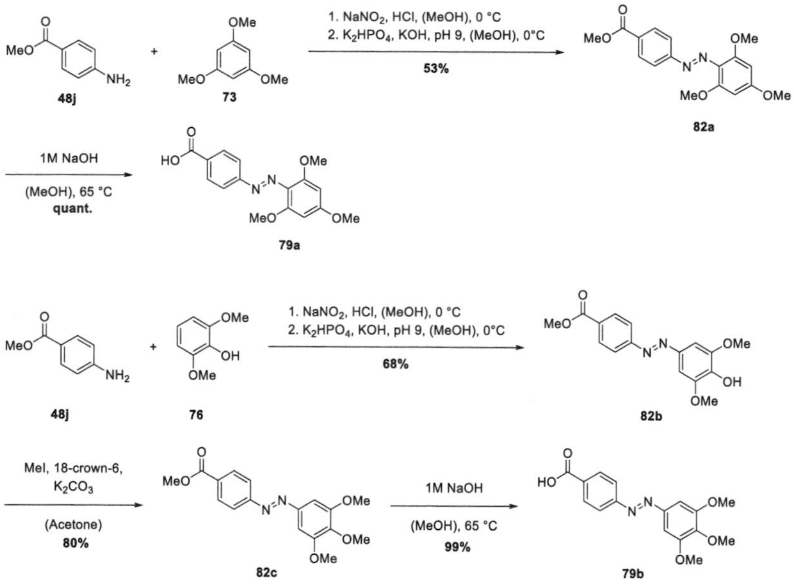

Scheme 34. Synthesis of azobenzenes **79a** and **79b**.

In addition to the azobenzene derivatives **79a-b** bearing methoxy substituents, another molecule design was investigated. In order to introduce more rigidity into the system, 1,3-benzodioxole was incorporated into the design. As shown in Scheme 35, intermediate **82d** was synthesized from an azo coupling reaction of methyl 4-aminobenzoate **48j** and catechol (**83**) with a yield of 20%. Subsequently, compound **82d** was reacted with diiodomethane in the presence of a base to yield the azobenzene derivative **82e** in a yield of 11%, as shown in Scheme 35.

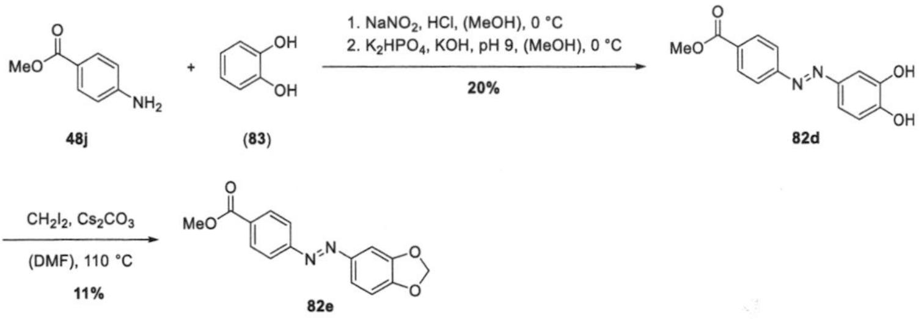

Scheme 35. Synthesis of azobenzene derivative **82e**.

Following the synthesis of the starting materials for the amide coupling, **Azo-Menthyl-*N*-amide-3,4,5OMe** (**77**) was synthesized as shown in Scheme 36 using HATU as a coupling reagent and Hünig's base. **Azo-Menthyl-*N*-amide-3,4,5OMe** (**77**) was obtained as an orange solid with a yield of 48%. The NOESY spectrum confirmed the stereochemistry and is shown in chapter 8.1, Figure 62.

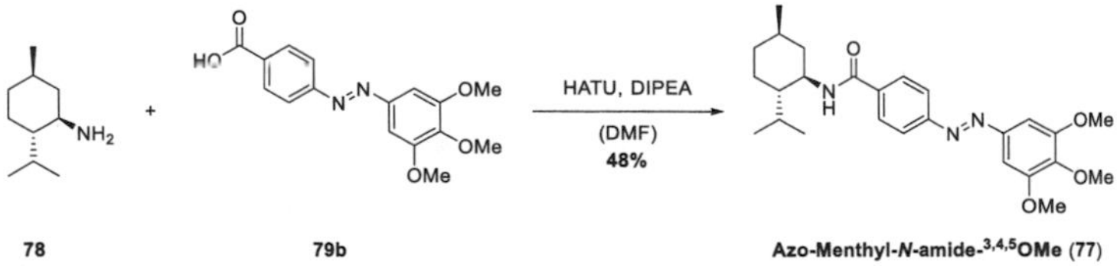

Scheme 36. Synthesis of **Azo-Menthyl-N-amide-3,4,5OMe** (**77**).

3.4.3. Azo-Menthyl-*O*-Ester Derivatives

With the respective carboxylic acid derivatives of azobenzene **79a-b** in hand, esterification with (−)-menthol using DCC as an activating reagent was conducted as displayed in Scheme 37. **Azo-menthyl-*O*-ester-2,4,6OMe** (**84a**) was synthesized with a yield of 28%, while **Azo-menthyl-*O*-ester-3,4,5OMe** (**84b**) was obtained with a yield of 48%. Both compounds were

126

obtained as orange-red oils. NOESY spectra confirmed the stereochemistry (chapter 8.1, Figure 63 and Figure 64).

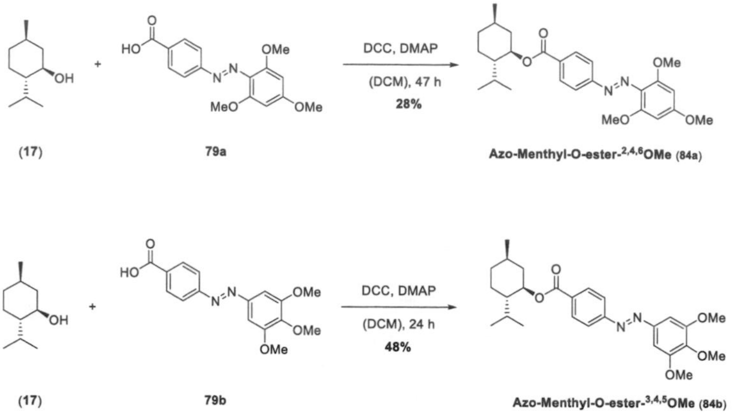

Scheme 37. Synthesis of azo-menthyl-O-ester derivatives **84a** and **84b**.

In summary, seven azo-menthol derivatives were synthesized. The synthesis of **Azo-Menthol (69)**, which Dr. David Konrad had established, was successfully conducted and was followed up by the derivatization of the azobenzene unit with methoxy substituents to obtain the regioisomers **Azo-Menthol-2,4,6OMe (70a)** and **Azo-Menthol-3,4,5OMe (70b)**, and using a diethylamino substituent which led to the formation of **Azo-Menthol-pNEt2 (70c)**. Furthermore, the N-amide isomer to **Azo-Menthol-3,4,5OMe (70b)**, **Azo-Menthyl-N-amide-3,4,5OMe (77)** was successfully obtained. Lastly, two ester derivatives, namely **Azo-Menthyl-O-ester-2,4,6OMe (84a)** and **Azo-Menthyl-O-ester-3,4,5OMe (84b)** were obtained. These compounds will be further investigated regarding the photophysical properties, especially concerning the photoswitching behavior, and in biological testing in collaboration with the Trauner group and other collaborators.

3.4.4. Azobenzenes – Absorption Behavior

After establishing the synthesis pathways, basic absorption spectra for some of the azobenzenes were obtained in acetonitrile or dichloromethane and are shown in Figure 45 and Table 17.

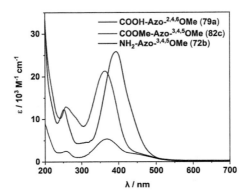

Figure 45. UV-vis spectroscopy data for azobenzenes **79a, 82c** and **72b** measured in acetonitrile.

COOH-Azo-^{2,4,6}OMe (79a) and **COOMe-Azo-^{3,4,5}OMe (82c)** both show absorption bands at around 258 nm and different bands at 367 and 363 nm, respectively. Although the absorption peaks are similar for both compounds, the extinction coefficients for the methyl ester **COOMe-Azo-^{3,4,5}OMe (82c)** are higher than for the carboxylic acid **COOH-Azo-^{2,4,6}OMe (79a)**. **NH$_2$-Azo-^{3,4,5}OMe (72b)** displays two absorption bands at 252 and 393 nm, while at 435 nm, there seems to be an underlying signal. In comparison to the two derivatives mentioned in the above paragraph, it shows a significantly increased extinction coefficient for the transition at 393 nm. Furthermore, the absorption band corresponding to this transition is red-shifted by 30 nm compared to the methyl ester substituted analog (Table 17).

Table 17. Summary of the UV-vis spectroscopy data for azobenzenes **79a, 82c** and **72b** measured in acetonitrile.

Entry	Compound	λ_{abs} (ε) [nm] ([10^3 M^{-1} cm^{-1}])
1	COOH-Azo-^{2,4,6}OMe (79a)	258 (2.50), 367 (5.39)
2	COOMe-Azo-^{3,4,5}OMe (82c)	258 (12.93), 363 (21.28)
3	NH$_2$-Azo-^{3,4,5}OMe (72b)	252 (12.32), 393 (25.84)

The data of the UV-vis measurements of **COOMe-Azo-^{2,4,6}OMe** (compound **82a**), **COOMe-Azo-^{3,5}OMe-^{4}OH** (compound **82b**), and **NO$_2$-Azo-^{2,4,6}OMe** (compound **74a**) was obtained from dichloromethane solutions and is shown in Figure 46 and Table 18.

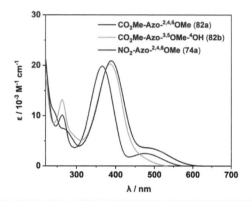

Figure 46. UV-vis spectroscopy data for azobenzene derivative **82a-b** and **74a**, measured in dichloromethane solution.

Compound **82a** showed absorption maxima and the corresponding extinction coefficients at 263 (10.11), 366 (19.83), and 478 nm (2.47·10^3 M^{-1} cm^{-1}), while compound **82b** displays absorption peaks at 263 and 386 nm. Lastly, compound **74a** showed absorption maxima at 242, 271, 389, and 498 nm with extinction coefficients of 10.87, 7.27, 20.89, and 3.41·10^3 M^{-1} cm^{-1}, respectively.

Table 18. Summary of the UV-vis spectroscopy data for azobenzene derivative **82a-b** and **74a**, measured in dichloromethane solution.

Entry	Compound	λ_{abs} (ε) [nm] ([10^3 M^{-1} cm^{-1}])
1	COOMe-Azo-^{2,4,6}OMe (82a)	263 (10.11), 366 (19.83), 478 (2.47)
2	COOMe-Azo-^{3,5}OMe-^{4}OH (82b)	263 (13.18), 386 (20.29)
3	NO$_2$-Azo-^{2,4,6}OMe (74a)	242 (10.87), 271 (7.27), 389 (20.89), 498 (3.41)

3.5. Photoswitchable Fluorophore Molecule Design

The photoisomerization of azobenzenes between its *trans*- and *cis*-isomer can be used to bring substituents on either side into spatial proximity. This can induce through-space charge-transfer (TSCT) of thermally activated delayed fluorescence (TADF) emitters.

In their *cis*-form, spatial π-π interaction between the electron-donating and –accepting moieties (donor and acceptor) would induce TADF. However, in their *trans*-form, the distance between the donor (HOMO) and acceptor (LUMO) without orbital overlap would result in switched-off luminescence as visualized in Scheme 38. This concept proposed a novel design of photoswitchable organic luminescent materials and was investigated using a model system. The model system featured a carbazole unit as the donor and a benzonitrile as an acceptor. The azobenzene acts as a bridging unit between the acceptor and donor and could be interpreted as a switchable spacer unit.

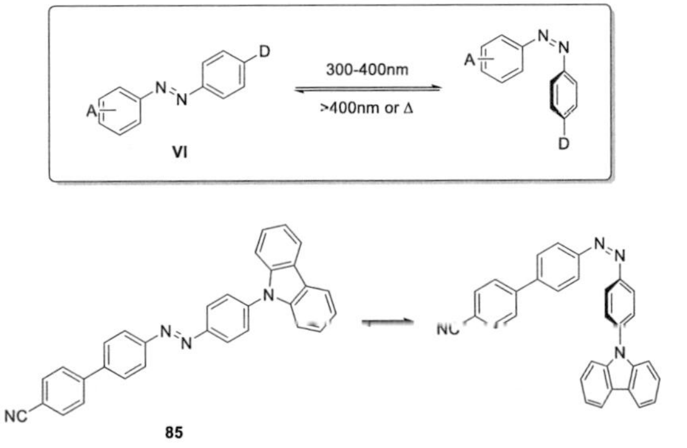

Scheme 38. Concept of photoswitchable TADF molecules and structure of the investigated model system 85.

Several routes for the synthesis of compound **85** were investigated. The target-leading synthesis route employed a BAEYER-MILLS reaction to form the azobenzenes core of **Br-Azo-F (86)** in a yield of 22%. Then, the donor was installed selectively on one side through a nucleophilic aromatic substitution to obtain compound **87** in a yield of 65%. Subsequently, the acceptor was installed *via* a BUCHWALD-HARTWIG cross-coupling reaction to give compound **88** as a yellow crystalline solid with a yield of 11%, as shown in Scheme 39.

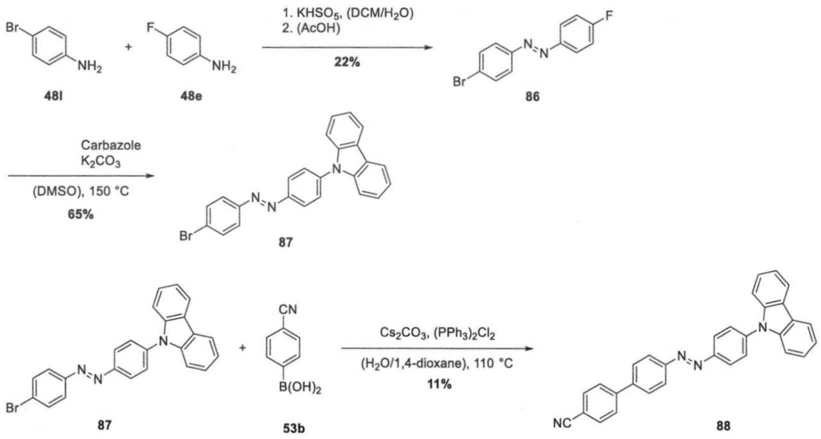

130

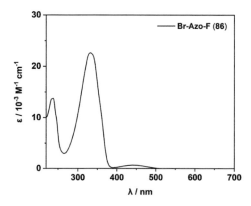

Scheme 39. Synthesis of the push-pull azobenzene compound 88.

In order to functionalize the building block-like intermediate 87, the synthesis of organoboron derivatives was attempted. However, neither the reaction of compound 87 with *n*-BuLi and trimethyl borate in THF, nor the reaction with bis(pinacolato)diboron in the presence of potassium acetate and Pd(PPh$_3$)$_2$Cl$_2$ in 1,4-dioxane lead to the formation of the desired boronic acid derivate or borate ester derivative, respectively. In conclusion, a model system for a photoswitchable donor-acceptor system for through-space induced charge-transfer was designed and successfully synthesized on a test scale.

Figure 47. Absorption and excitation spectra of **Br-Azo-F** (86) measured in dichloromethane.

While the in-depth photophysical characterization of these compounds will follow in collaboration with partners, the absorption spectrum of **Br-Azo-F (86)** was measured in DCM and is shown in Figure 47. It shows two distinct absorption bands with corresponding extinction coefficients in parenthesis at 237 (13.75) and 333 nm (22.61 10^3 M^{-1} cm^{-1}), as well as a broad band around 443 nm (0.65 10^3 M^{-1} cm^{-1}).

4. Conclusion and Outlook

In conclusion, this work encompasses the synthesis of light-harvesting organic molecules. First, the design and synthesis of dibenzo[a,c]phenazine (BP)-based TADF emitters were established. Then, two more electron-accepting systems, namely pyrrolo[3,4-f]isoindole-1,3,5,7(2H,6H)-tetraone (PIT) and 1H-pyrrolo[3,4-b]quinoxaline-1,3(2H)-dione (PQD), were designed and synthesized. The thereby obtained acceptors and (poly-)donor-acceptor conjugates were investigated regarding their electrochemical properties as well as their absorption and emission behavior. After developing luminescent molecules that harvest light of a certain energy to emit light of a specific color, molecules that translate light into conformational changes were investigated. The syntheses of **Azo-Menthol (69)** derivatives that show red-shifted absorption were conducted and lastly, the photo-induced conformational change was envisaged to control the properties of molecules that emit *via* through-space induced charge transfer. In the following subsections, the results and highlights of each project are summarized and future investigations are outlined.

4.1. Dibenzo[a,c]phenazine (BP)-Based Molecules

The investigated BP-based molecule designs are the following: D-A, 2D-A-D, 2D-A-F, 2D-A-OtBu, and 2D-A-R, and the thereby obtained molecule structures are summarized in Figure 48 and Figure 49.

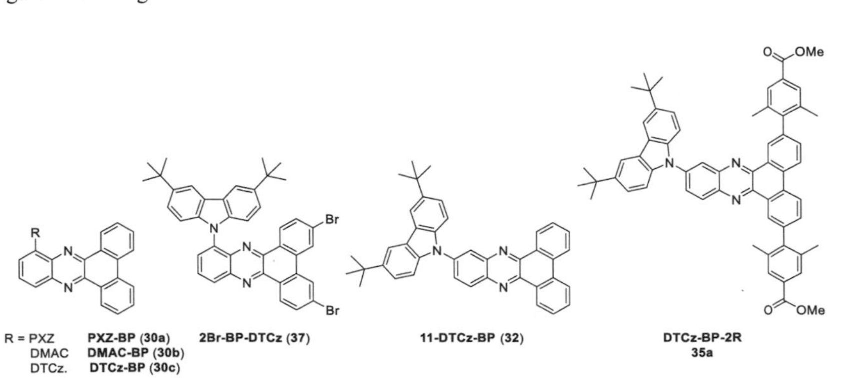

R = PXZ   **PXZ-BP (30a)**   **2Br-BP-DTCz (37)**   **11-DTCz-BP (32)**   **DTCz-BP-2R**
DMAC   **DMAC-BP (30b)**   **35a**
DTCz.   **DTCz-BP (30c)**

Figure 48. D-A-type BP-based structures.

For the D-A-type BP-based emitter, an investigation of the influence of different donor units was conducted by using three different electron-donating units, namely DTCz, DMAC, and PXZ, to yield **PXZ-BP (30a)**, **DMAC-BP (30b)**, and **DTCz-BP (30c)** in moderate to good yields. Furthermore, **11-DTCz-BP (32)** was custom-designed to fit into the pore of a MIL-68(In) MOF. Lastly, a D-A-type dibenzo[*a,c*]phenazine-based MOF linker precursor compound **35a** was synthesized. The thereby obtained substrates were investigated employing TD-DFT to gain insight into their electronic natures as well as the suitability of this design for thermally activated delayed fluorescence. The trend of the donor strength could be observed for **PXZ-BP (30a)**, **DMAC-BP (30b)** and **DTCz-BP (30c)**, as a shallower HOMO level correlated to an increased donor strength. The time-dependent DFT calculation for the excited states revealed a minuscule ΔE_{ST} for **PXZ-BP (30a)** and **DMAC-BP (30b)**, but **DTCz-BP (30c)** displayed a larger value due to a bigger FMO overlap. In addition to investigating the donor strength effect on the D-BP-type molecules, the connectivity of the donor moiety to the BP acceptor was varied. **11-DTCz-BP (32)**, an isomer to **DTCz-BP (30c)**, was found to depict a less shallow HOMO and an increased LUMO level, as well as a higher ΔE_{ST} because of an increased FMO overlap. Subsequently, the absorption and emission behavior of the D-A-type fluorophores were investigated. **PXZ-BP (30a)** showed the furthest red-shifted emission with a maximum wavelength of 716 nm, followed by the deep-red emitting **DMAC-BP (30b)** (648 nm) and the green-emitting **DTCz-BP (30c)** (561 nm). The emission wavelength of the MOF-linker precursor compound **35a** was blue-shifted to 518 nm in comparison to **DTCz-BP (30c)**. The absorption spectra for **DTCz-BP (30c)** and **DTCz-BP-2R (35a)** furthermore indicate that the π-backbone of the latter has a significant impact on the electronic structure of the molecule.

After establishing the D-A-type fluorophores based on dibenzo[*a,c*]phenazine, a (poly-donor)-acceptor approach was applied to obtain 2D-A-D, 2D-A-F, 2D-A-O*t*Bu, and 2D-A-R-type substrates, which are summarized in Figure 49. The 2D-A-D molecule design aimed to investigate adding more donors for a red-shifted emission. The experiments relating to the 2D-A-F, 2D-A-O*t*Bu, and 2D-A-R-type substrates were aimed at elucidating the effect of acceptor modification through different types of substituents.

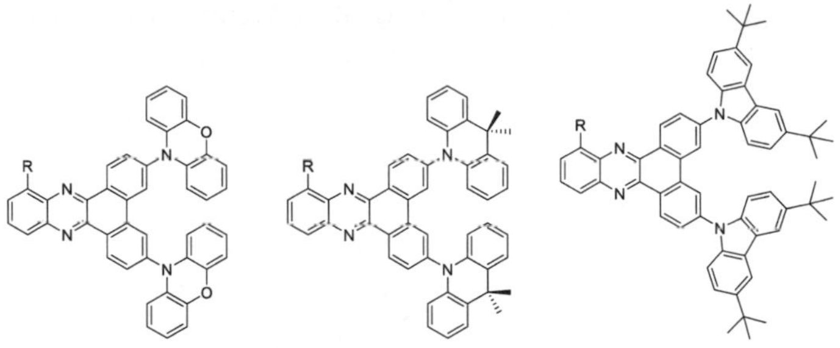

R = F   **2PXZ-BP-F (39a)**
O*t*Bu **2PXZ-BP-O*t*Bu (40a)**
DTCz **2PXZ-BP-DTCz (38)**

R = F   **2DMAC-BP-F (39b)**
O*t*Bu **2DMAC-BP-OtBu (40b)**

R = F   **2DTCz-BP-F (39c)**
O*t*Bu **2DTCz-BP-O*t*Bu (40c)**
DTCz **2DTCz-BP-DTCz (39d)**

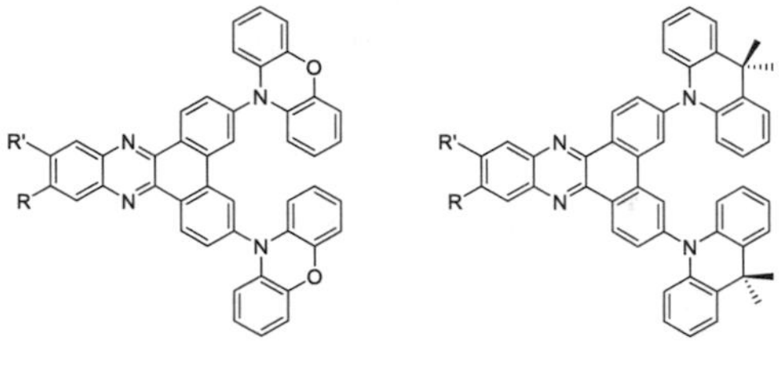

R = R' = F   **2PXZ-BP-2F (43c)**

Me   **2PXZ-BP-2Me (43d)**

R = H, R' = COPh **2PXZ-BP-COPh (43e)**

R = R' = F   **2DMAC-BP-2F (44c)**
Me   **2DMAC-BP-2Me (44d)**

R = H, R' = COPh **2DMAC-BP-COPh (44e)**
SO₃H **2DMAC-BP-SO₃H (44f)**

Figure 49. 2D-A-D-, 2D-A-F, 2D-A-O*t*Bu, and 2D-A-R-type structures based on the BP acceptor.

Following the synthetic access to the D-A-type substrates, the synthesis of 2D-BP-DTCz-type structures was initiated with the attachment of the donor moiety to **2Br-BP-F (36)** *via* S$_N$Ar. Although this was attempted with the PXZ, DMAC and DTCz donor, the S$_N$Ar was only successful with the DTCz donor, leading to the synthesis of **2Br-BP-DTCz (37)**. From here, **2PXZ-BP-DTCz (38)** was synthesized in a directed fashion. The second 2D-A-D-type substrate, **2DTCz-BP-DTCz (39d)** was obtained as a side product of a BUCHWALD-HARTWIG amination of the fluorinated precursor **2Br-BP-F (36)**. The synthesis of **2DTCz-BP-F (39c)** could be realized by increasing the amount of P(*t*Bu)₃ ligand and temperature. The PXZ- and DMAC-employing analogs, **2PXZ-BP-F (39a)** and **2DMAC-BP-F (39b)** were synthesized and completed the three-part series of the 2D-BP-F molecules. A second side reaction to the abovementioned reaction led to the formation of **2DTCz-BP-O*t*Bu (40c)** *via* Pd-mediated aryl ether formation. Based on this finding, the 2D-BP-O*t*Bu series containing **2PXZ-BP-O*t*Bu (40a)**, **2DMAC-BP-O*t*Bu (40b)** and **2DTCz-BP-O*t*Bu (40c)** was obtained. Lastly, another modular synthesis strategy was established by which the acceptor core could be easily derivatized using commercially available 1,2-diaminobenzene derivatives. The 2PXZ-BP-R and 2DMAC-BP-R series showing two fluorine substituents (**43c**) and (**44c**), two methyl substituents (**43d**) and (**44d**), or a benzoyl unit (**43e**) and (**44e**) attached to the BP core, and **2DMAC-BP-SO₃H** (**44f**) were synthesized with minimal synthetic effort.

Additionally, the synthesized 2D-BP-D, 2D-BP-F, 2D-BP-O*t*Bu and 2D-BP-R-type molecules were computed employing TD-DFT, and cyclic voltammetry measurements were conducted. The DFT calculation for the 2D-A-D-type structures showed that the third donor (DTCz) does not actively contribute to HOMO or LUMO. Thus, a leaner molecule design was considered effective in yielding TADF emitters demonstrating strong ICT, and the 2D-A-R-type substrates were investigated. The investigation of the 2D-A-R-type compounds showed that the LUMO energies corresponded to the electron-withdrawing nature of the substituents. Thereby, the substitution pattern can be adjusted to fine-tune the acceptor strength of the dibenzo[*a,c*]phenazine acceptor core. The fluorescence spectroscopy data showed that **2PXZ-BP-COPh** (**43e**) showed the furthest red-shifted emission with a maximum emission wavelength of 680 nm. Furthermore, OLED devices built with the 2D-BP-F series (**39a-c**) showed donor-strength-dependent green to orange-red emission with EQEs of up to 21.8% for the OLED device featuring **2DMAC-BP-F** (**39b**), and a low efficiency roll-off for the OLED device featuring **2PXZ-BP-F** (**39a**).

In developing novel acceptor systems for TADF emitters, efforts have been made to develop the BP acceptor design further. The novel acceptor designs include the addition of auxiliary acceptors, more heteroatoms, and/or an enlarged π-conjugated system. [163-166] Future investigations of the BP-based emitters will focus on incorporating organelle-targeting molecule designs. The synthesis from compound **90** onwards that has not been completed and is therefore not included in this work is shown in Figure 50. Inspiration was drawn from literature-reported emitters for TRLI **AI-Cz-MT** (**8a**) and **AI-Cz-LT** (**8b**) (Figure 4).[47]

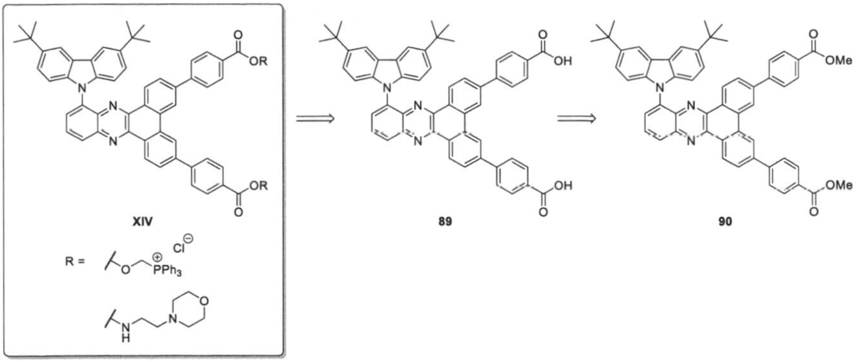

Figure 50. The future project focus on BP-based, organelle-targeted emitters **XIV**.

Lastly, the BP-based MOF linkers were explored by synthesizing compound **35a**. The derivatives bearing the stronger donors PXZ and DMAC, namely compound **35b** and **35c** (Figure 51), were also synthesized, but not shown in this work as the reactions were only carried out on a test scale and the purity was not satisfactory. Future investigations ought to be carried out to upscale the syntheses and carry out further characterization as well as hydrolysis of the ester moiety to obtain the carboxylic acid derivatives.

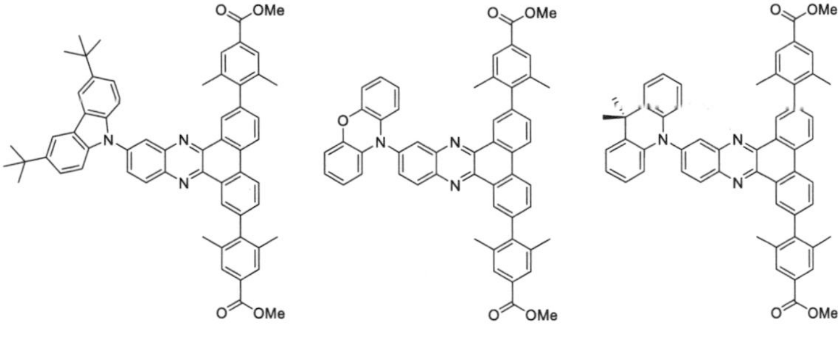

Figure 51. Molecular structures of BP-based MOF linker precursors **35a-c**.

4.2. Pyrrolo[3,4-*f*]isoindole-1,3,5,7(2*H*,6*H*)-tetraone (PIT)-Based Molecules

The pyrrolo[3,4-*f*]isoindole-1,3,5,7(2*H*,6*H*)-tetraone (PIT)-based TADF molecule design was developed and investigated in this work. The molecule design originated from the extension of phtalimide-based TADF molecules, combined with inspiration drawn from a fluorescent anthracene derivative reported by Adachi and coworkers.[149] The synthesis of PIT-based molecules underwent a continuous re-adjustment of the molecule designs as establishing the synthesis route presented challenges. First, D-A-type structures were targeted. However, as the BUCHWALD-HARTWIG amination reaction was unsuccessful in leading to the target structure, the synthesis focus shifted to D-π-A-π-D-type structures, whose synthesis was attempted *via* SUZUKI-MIYAURA cross-coupling. Although an extensive study of reaction conditions was investigated, the synthesis of D-π-A-π-D-type PIT-based molecules could not be realized. Thus, a precursor for a STILLE cross-coupling was prepared, and the cross-coupling reaction was conducted. This, too, was not target-leading. As these cross-coupling reactions were not expedient, the initial starting point of the molecule design was reflected upon. Adachi and co-workers had connected their fluorescent anthracene derivative core *via* an alkyne bond. Thus, SONOGASHIRA cross-coupling was applied to the PIT system, and finally, the target compound **PXZ-π-#-PIT-#-π-PXZ** (**57**) (Figure 52) was obtained.

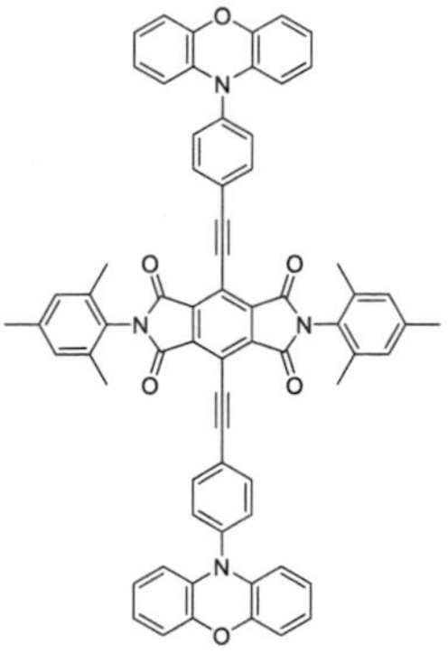

PXZ-π-#-PIT-#-π-PXZ
(57)

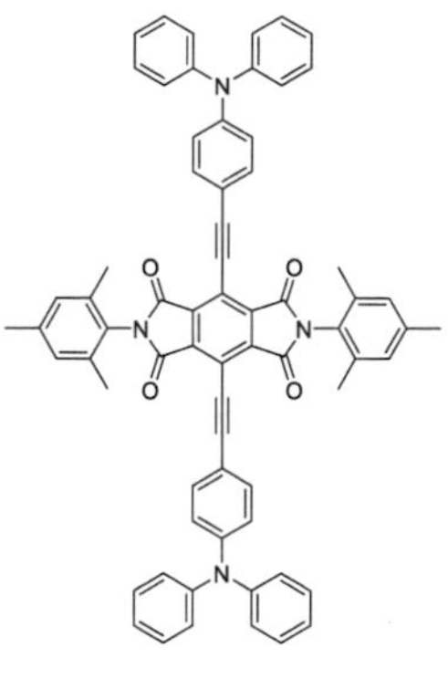

TPA-π-#-PIT-#-π-TPA
(91)

Figure 52. Structure of **PXZ-π-#-PIT-#-π-PXZ** (**57**) and **TPA-π-#-PIT-#-π-TPA** (**91**).

Interestingly, this compound did not show luminescence at room temperature and only faint luminescence at 77 K in toluene solution. When investigated by UV-vis and fluorescence spectroscopy, compound (**57**) showed a red-shifted ICT absorption band compared to literature reported compounds.[154-155] Furthermore, it showed low-intensity fluorescence with emission wavelength maxima at 414 and 565 nm. The geometry optimization by means of DFT calculation showed that the HOMO was mainly localized on the PXZ moiety, while the LUMO was spread out over the PIT acceptor core. The TD-DFT calculation results of the excited states furthermore depicted a small ΔE_{ST} of 0.02 eV. In addition, the compound was investigated by cyclic voltammetry, and experiment values for the HOMO and LUMO energy levels were obtained. Although compound (**57**) did not show strong fluorescence as desired, the TD-DFT calculation result demonstrates the promising nature of this acceptor for TADF materials. In future investigations, **TPA-π-#-PIT-#-π-TPA** (**91**) featuring TPA donors is to be synthesized and characterized.

4.3. 1*H*-Pyrrolo[3,4-*b*]quinoxaline-1,3(2*H*)-dione (PQD)-Based Molecules

Following the establishment of the PIT acceptor, the phthalimide-based acceptor design was extended and led to the design of 1*H*-pyrrolo[3,4-*b*]quinoxaline-1,3(2*H*)-dione (PQD)-based acceptors. Here, an acceptor precursor design was extensively investigated starting from a literature-known synthesis of a dicarboxylated quinoxaline derivative. By optimizing the reaction conditions, the synthesis route to the imide formation was established and various amines were employed to demonstrate a glimpse into the scope of this reaction (**64a-c, 64e-g,** Figure 53). One donor-acceptor conjugate **Br-PQD-TPA** (**64d**, Figure 54) was obtained *via* this synthesis route.

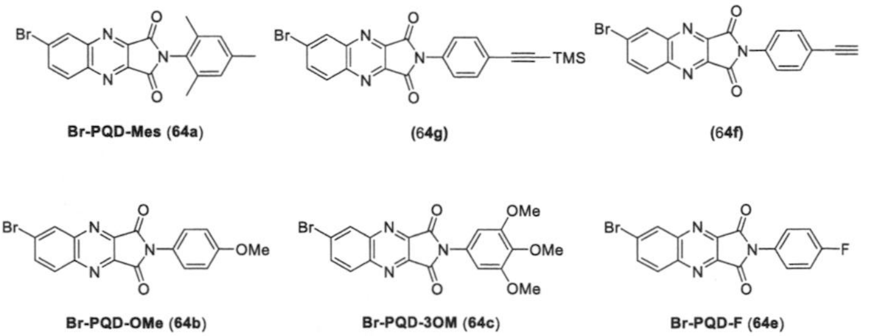

Br-PQD-Mes (64a)　　　(**64g**)　　　(**64f**)

Br-PQD-OMe (64b)　　**Br-PQD-3OM (64c)**　　**Br-PQD-F (64e)**

Figure 53. Molecular structures of PQD acceptor derivatives (**64a-c**) and (**64e-g**) serving as precursors for D-A and D-π-A-type conjugates.

To access more D-A and D-π-A-type structures, the obtained Br-PQD-R-type molecules were subjected to cross-coupling reactions. This led to the synthesis of **PXZ-PQD-Mes (67)**, **PXZ-π-PQD-Mes (68)**, as well as **TPA-PQD-Mes (66a)** and **TPA-PQD-OMe (66b)**, which are shown in Figure 54.

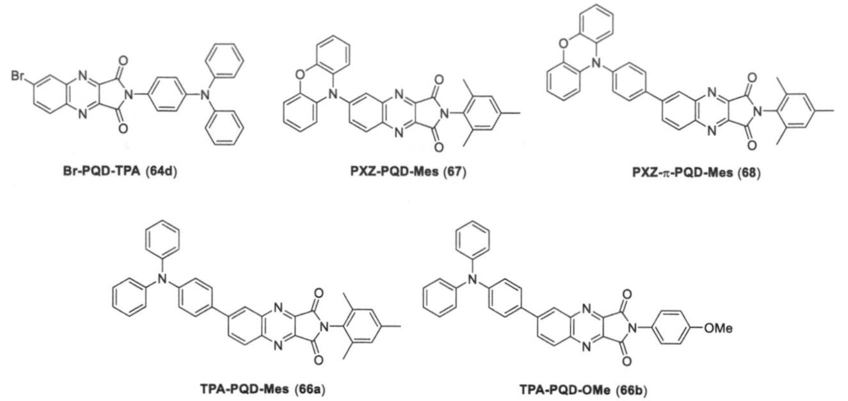

Br-PQD-TPA (64d) PXZ-PQD-Mes (67) PXZ-π-PQD-Mes (68)

TPA-PQD-Mes (66a) TPA-PQD-OMe (66b)

Figure 54. Molecular structures of the D-A- and D-π-A-type PQD-based compounds.

The acceptor precursors as well as D-A- and D-π-A-type PQD-based compounds were investigated by means of cyclic voltammetry and in the case of the latter also by employing TD-DFT calculation. The acceptor strengths of the PQD acceptors **(64a-c)** and **(64e-g)** shown in Figure 53 were determined by means of CV, and the measurements revealed that the acceptor strength could be easily modified by the nature of the substitution pattern on the aryl ring attached at the imide position. **Br-PQD-F (64e)** demonstrated the lowest LUMO energy and is therefore the strongest acceptor amongst the investigated Br-PQD-R-type compounds, while **Br-PQD-OMe (64b)** depicted the least stabilized LUMO level and is thus the weakest in this series. The CV results furthermore gave information about the electrochemical stability of the Br-PQD-R series. While **Br-PQD-Mes (64a)**, **Br-PQD-F (64e)** and **Br-PQD-3OMe (64c)** showed reversible reduction processes and good electrochemical stability, compound **(64f)** showed pseudo-reversibly and **Br-PQD-OMe (64b)** an irreversible reduction process. UV-vis and fluorescence spectroscopy measurements showed that except for **Br-PQD-F (64e)**, the absorption and emission maxima were bathochromically shifted depending on the number of substituents and the electron density they contribute. Compound **(64e)** showed two emission wavelengths that could be selectively addressed by adjusting the excitation wavelength.

The CV measurements provided experimental values for the HOMO and LUMO energies for the D-A-type compounds **TPA-PQD-Mes (66a)** and **TPA-PQD-OMe (66b)**, displayed electrochemical stability of both compounds and showed that the substituent variation at the imide position has a negligible impact on the HOMO/LUMO energies in this case. Furthermore, the experimentally obtained values matched the theoretically calculated values by TD-DFT calculations. The TD-DFT calculation for **PXZ-π-PQD-Mes (68)** of the excited states indicated that the ΔE_{ST} amongst **TPA-PQD-Mes (66a)**, **TPA-PQD-OMe (66b)** and **PXZ-π-PQD-Mes (68)** is smallest for the latter. Although the minimal ΔE_{ST} is a valuable indicator for efficient RISC and TADF, **PXZ-π-PQD-Mes (68)** only showed weak emission at room temperature with a wavelength maximum of 677 nm and stronger yellow-orange luminescence at 77 K. **PXZ-PQD-Mes (67)** on the other hand showed extremely low emission intensity. The PXZ-bearing molecules **(67)** and **(68)** demonstrated in parts that the molecule design successfully produced deep-red emitting luminophores. However, it lacked in terms of delivering efficient luminescence. Changing the donor moiety from PXZ to TPA yielded an over 85-fold increase in the extinction coefficient of the absorption band assigned to ICT for **TPA-PQD-Mes (66a)** in comparison to **PXZ-PQD-Mes (67)**. Both **TPA-PQD-Mes (66a)** and **TPA-PQD-OMe (66b)** showed orange-red emission with wavelength maxima at 581 and 587 nm, respectively. Finally, **Br-PQD-TPA (64d)**, which exhibited different donor-to-acceptor connectivity, displayed a significantly red-shifted emission with a maximum wavelength of 662 nm. As the different connectivity reduced the overlap of the FMOs *via* the imide unit, it was reasoned that the acceptor core was strengthened.

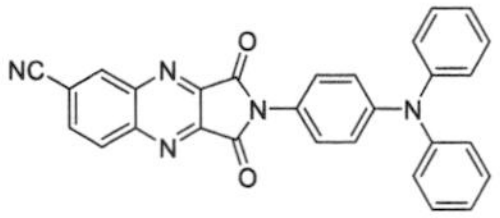

92

Figure 55. Molecular design of PQD-based emitter **92**.

Building on this finding, future investigations regarding PQD-based emitters will focus on synthesizing compound **92** (Figure 55) to achieve an even stronger red-shift and high-efficiency emitter in the fashion of Bronstein and coworkers.[121]

142

4.4. Red-Shifted Azo-Menthol Derivatives

The synthesis of derivatives of the **Azo-Menthol** (**69**) structure that Dr. David Konrad developed, was successfully achieved by synthesizing six azo-menthyl compounds shown in Figure 56. The derivatives displayed various substituents of the azobenzene substructures, for which the synthesis route was successfully established. Several connectivities between the menthyl unit and the azobenzene moiety were explored and thereby, Azo-Menthyl-*C*-amides (**70a-c**), Azo-Menthyl-*N*-amides (**77**) and Azo-Menthol-*O*-esters (**84a-b**) were obtained. The photophysical characterization analyzing the photoswitching behavior of these compounds will be conducted in corporation with the Trauner group and is ongoing. Furthermore, these compounds will be tested in biological testing concerning their behavior as TRPM8-addressing substrates.

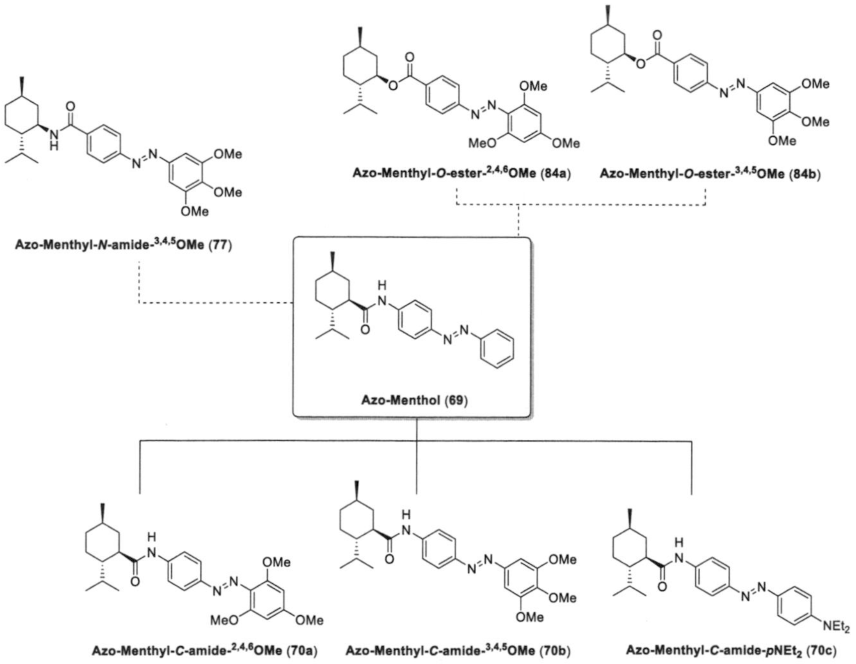

Figure 56. Molecular structures of the Azo-Menthol derivatives which can be categorized as Azo-Menthyl-*C*-amides, Azo-Menthyl-*N*-amides, and Azo-Menthyl-*O*-esters.

4.5. Photo-switchable Fluorophore Molecule Design

Lastly, the underlying design concept of photoswitchable azobenzenes and molecular structures showing thermally activated delayed fluorescence was combined to establish the synthesis of the push-pull/donor-acceptor system **88** that would serve as a model substrate to investigate switch-on and -off through-space induced charge transfer systems. Future investigations will concern the development of other push-pull azobenzenes **XV** with triazine acceptors and stronger donors such as PXZ and DMAC.

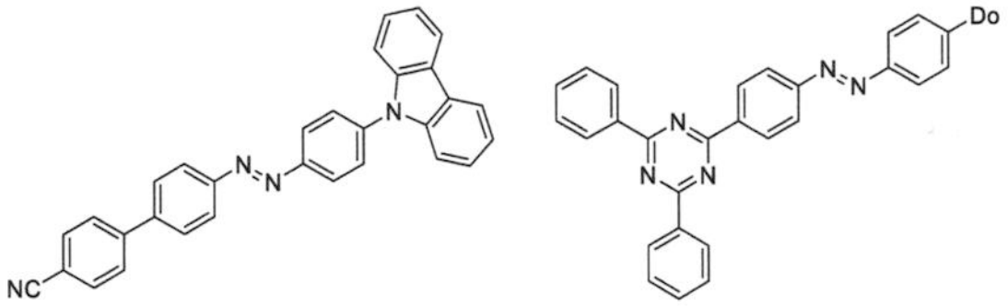

88   **XV**, Do = PXZ, DMAC

Figure 57. Structure of push-pull azobenzenes **88** and **XV** featuring donor and acceptor moieties.

5. Experimental Section

5.1. General Remarks

Materials and Methods

Solvents, reagents, and chemicals were purchased from abcr, Fluorochem, Sigma Aldrich, Merck, TCI, and Chempur and used without further purification. Air- and moisture-sensitive reactions were carried out under argon atmosphere in sealable vials or flame-dried flasks using standard Schlenk techniques. Liquids were added with a stainless-steel cannula, and solids were added in a powdered shape. Solvents were evaporated under reduced pressure at 45 °C using a rotary evaporator. For solvent mixtures, each solvent was measured volumetrically. Column chromatography was performed using Merck silica gel 60 (0.040 × 0.063 mm, 230–400 mesh ASTM) and quartz sand (glowed and purified with hydrochloric acid). Flash chromatography was carried out either using a CombiFlash Rf+ device by Teledyne Isco with reuseable RediSep Rf cartridges by VERTEX Technics S.L or a PureC-815 Flash device by Büchi in combination with the following columns: Büchi FlashPure EcoFlex (4 g, Silica 50 μm irregular), Interchim PuriFlash® (40 g, Silica HC, 50 μm) or Interchim PuriFlash® (12 g, Silica HP, 50 μm).

Reaction Monitoring

Routine monitoring of reactions was performed using silica gel coated aluminum plates (Merck, silica gel 60, F254) analyzed under UV light at 254 and 365 nm. Solvent mixtures are understood as v/v.

Nuclear Magnetic Resonance Spectroscopy (NMR)

^{1}H NMR spectra were recorded on Bruker Avance 400 (400 MHz) and Bruker Avance DRX 500 (500 MHz) spectrometers. Chemical shifts are given in parts per million (δ/ppm), downfield from tetramethylsilane (TMS), and are referenced to chloroform (7.26 ppm), tetrahydrofuran (1.73 ppm), dimethylsulfoxide (2.50 ppm), dichloromethane (5.32 ppm) or benzene (7.16 ppm) as internal standard. All coupling constants are absolute values, and J values are expressed in Hertz (Hz). The description of signals includes: s = singlet, bs = broad singlet, d = doublet, t = triplet, dd = doublet of doublets, ddd = doublet of doublet of doublets, dt = doublet of triplets, q = quartet, quin = quintet, sxt = sextet, sept = septet, m = multiplet.

The spectra were analyzed according to the first order. ^{13}C NMR spectra were recorded on Bruker Avance 400 (100 MHz) and Bruker Avance DRX 500 (125 MHz) spectrometers. Chemical shifts are expressed in parts per million (δ/ppm) downfield from tetramethylsilane (TMS) and are referenced to chloroform (77.16 ppm), tetrahydrofuran (67.21 ppm), dimethylsulfoxide (39.52 ppm), dichloromethane (53.84 ppm) or benzene (128.06 ppm) as internal standard. ^{19}F NMR were recorded on Bruker Avance 400 (376 MHz) and the ^{11}B spectra (128 MHz). Chemical shifts are stated in parts per million (δ/ppm).

Mass Spectrometry (MS)

Electron ionization (EI) and fast atom bombardment (FAB) experiments were conducted using a Finnigan MAT 90 (70 eV) instrument, with 3-nitrobenzyl alcohol (3-NBA) as matrix and reference for high resolution. The ESI experiments were obtained using a Q-Exactive (Orbitrap) mass spectrometer by Thermo Fisher Scientific equipped with a HESI II probe to record high resolution. For the interpretation of the spectra, molecular peaks [M]$^+$, peaks of protonated molecules [M+H]$^+$, and characteristic fragment peaks are indicated with their mass-to-charge ratio (m/z) and their intensity in percent, relative to the base peak (100%) is given. In the case of high-resolution measurements, the tolerated error is 0.0005 m/z.

Infrared Spectroscopy (IR)

The infrared spectra were recorded with a Bruker, Alpha P instrument. All samples were measured by attenuated total reflection (ATR). The positions of the absorption bands are given in wavenumbers $\bar{\upsilon}$ in cm^{-1} and were measured in the range from 3600 cm^{-1} to 500 cm^{-1}. Characterization of the absorption bands was done in dependence on the transmission strength with the following abbreviations: vs (very strong, 100–90%), s (strong, 89–70%), m (medium, 59–40%), w (weak, 39–10%), vw (very weak, 0–9%).

Elemental Analysis (EA)

The elemental analysis measurements were carried out using an Elementar Vario MICRO instrument. Furthermore, the weight scale that was used was a Sartorius M2P. The calculated and found percentage by mass values for C, H, N and S are declared in fractions of 100%.

Melting Point (m.p.)

Melting points were measured using an OptiMelt MPA100 device by Stanford Research System.

Cyclic Voltammetry (CV)

Cyclic voltammetry experiments were performed with a Gamry Interface 1010B in a three electrodes electrochemical cell. The electrochemical cell was equipped with a glassy carbon (GC) working electrode, $Ag/AgNO_3$ reference electrode, and a Pt wire as the auxiliary electrode.

The experiments were performed in Ar-saturated dichloromethane or *N,N*-dimethylformamide containing 0.1 M [nBu_4N][PF_6] as the electrolyte at a scan rate of 100 mV s^{-1}. The concentration of the investigated compounds was 2 mM. Ferrocene (Fc) was added after each experiment as an internal standard, according to the IUPAC recommendation.[167]

The redox properties are reported versus the SCE couple according to the following equation:[168]

$$E_{ox/red} [V \text{ vs. SCE}] = E_{ox/red} [V \text{ vs. Fc/Fc}^+] + 0.46$$

The HOMO and LUMO energies were determined using $E_{HOMO/LUMO} = -(E_{ox/red}+4.8)$ eV, where E_{ox} and E_{red} represent the anodic and cathodic peak potentials, respectively, versus Fc/Fc$^+$.[169-170]

UV-vis Spectroscopy

The absorbance was recorded using an Analytik Jena Specord 50/plus or a Perkin-Elmer Lambda 650 spectrometer.

Fluorescence Spectroscopy

The emission was measured on a Horiba Scientific fluoromax-4 spectrofluorometer or an Agilent Cary Eclipse fluorescent spectrometer, both equipped with a Czerny-Turner-type monochromator and an R928P PMT detector.

5.2. Synthesis Procedures and Analytical Data

5.2.1. Dibenzo[*a,c*]phenazine (BP)-Based Molecule Design

3,6-Di(10*H*-phenoxazin-10-yl)phenanthrene-9,10-dione (28c)

A sealable vial was charged with 3',6'-di(10*H*-phenoxazin-10-yl) dispiro [[1,3] dioxolane-2,9'-phenanthrene-10',2''-[1,3] dioxolane] (602 mg, 914 µmol, 1.00 equiv.). Then, acetic acid (5.00 mL) was added and the reaction mixture was stirred at 120 °C and stirred for 5 d. Subsequently, all volatile components were removed under reduced pressure. The residue was dissolved in ethyl acetate and extracted with NaHCO$_3$ twice before being washed with brine. The organic phase was dried over Na$_2$SO$_4$ and concentrated under reduced pressure. The crude mixture was further purified employing column chromatography (cyclohexane : dichloromethane = 2:1). The product (434 mg, 761 µmol, 83% yield) was obtained as a dark red solid.

^{1}H NMR (400 MHz, CDCl$_3$) δ = 8.46 (d, ^3J = 8.3 Hz, 2H, 1-H), 7.96 (d, ^4J = 1.9 Hz, 2H, 4-H), 7.54 (dd, ^3J = 8.3 Hz, ^4J = 1.9 Hz, 2H, 2-H), 6.79–6.68 (m, 8H, 4'-H, 5'-H), 6.65 (ddd, ^3J = 8.4 Hz, ^3J = 6.9 Hz, ^4J = 2.2 Hz, 4H, 3'-H), 6.12 (dd, ^3J = 8.4 Hz, 1.3, 4H, 2'-H) ppm.

^{13}C NMR (101 MHz, CDCl$_3$) δ = 178.99 (2C, C=O), 147.33 (2C, C-3), 144.55 (4C, C-6'), 138.19 (2C, C-5), 133.74 (2C, C-1), 133.05 (4C, C-1'), 132.12 (2C, C-2), 130.32 (2C, C-6), 126.17 (2C, C-4), 123.59 (4C, C-3'), 122.80 (4C, C-4'), 116.32 (4C, C-5'), 113.94 (4C, C-2') ppm.

MS (FAB, 3-NBA), m/z (%): 571 (100) [M]$^+$.

HRMS–FAB *(m/z)*: Calc. for [C$_{38}$H$_{23}$O$_4$N$_2$]$^+$: 571.1652; found: 571.1652.

IR (ATR, ṽ) = 3061 (w), 1673 (m), 1588 (s), 1482 (vs), 1460 (s), 1401 (w), 1320 (vs), 1290 (s), 1269 (vs), 1259 (vs), 1197 (s), 925 (m), 741 (vs), 730 (vs), 700 (s), 629 (s) cm^{-1}.

R$_f$ = 0.48 (Cyclohexane : ethyl acetate = 4:1).

3,6-bis(9,9-dimethylacridin-10(9*H*)-yl)phenanthrene-9,10-dione (28d)

A sealable vial was charged with 3',6'-bis(9,9-dimethylacridin-10(9*H*)-yl) dispiro [[1,3]dioxolane-2,9'-phenanthrene-10',2''-[1,3] dioxolane] (899 mg, 1.26 mmol, 1.00 equiv.). Then, acetic acid (7.00 mL) was added, and the reaction mixture was stirred at 120 °C for 5 d. Subsequently, all volatile components were removed under reduced pressure. The residue was dissolved in ethyl acetate and extracted with NaHCO$_3$ twice before being washed with brine. The organic phase is dried over Na$_2$SO$_4$ and concentrated under reduced pressure. The crude mixture was further purified by means of column chromatography (cyclohexane : ethyl acetate = 20:1 → cyclohexane : dichloromethane = 1:1). The product (620 mg, 996 μmol, 79% yield) was obtained as a dark red solid.

¹H NMR (400 MHz, CDCl$_3$) δ = 8.38 (d, ^3J = 8.4 Hz, 2H, 2-H), 7.89 (d, ^4J = 2.0 Hz, 2H, 5-H), 7.54–7.44 (m, 6H, 3-H), 7.09–6.95 (m, 8H, 3'-H), 6.64–6.57 (m, 4H, 2'-H), 1.63 (s, 12H, C*H*$_3$) ppm.

¹³C NMR (101 MHz, CDCl$_3$) δ = 179.55 (2C, C=O), 149.74 (2C, C-4), 140.46 (4C, C-1'), 138.53 (2C, C-6), 133.93 (4C, C-6'), 133.32 (2C, C-2), 129.55 (2C, C-1), 129.44 (2C, C-3), 126.80 (4C, C-3'), 125.73 (4C, C-5'), 123.42 (2C, C-5), 122.75 (4C, C-4'), 116.84 (4C, C-2'), 36.87 (2C, C$_q$), 30.41 (4C, C*H*$_3$) ppm.

MS (FAB, 3-NBA), m/z (%):623 (99) [M+H]$^+$, 622 (24) [M]$^+$.

HRMS–FAB *(m/z)*: Calc. for [C$_{44}$H$_{35}$O$_2$N$_2$]$^+$: 623.2693; found: 623.2693.

IR (ATR, ṽ) = 3067 (w), 2965 (w), 2952 (w), 2921 (w), 2856 (w), 1670 (w), 1585 (vs), 1499 (w), 1470 (s), 1446 (vs), 1315 (vs), 1288 (m), 1261 (vs), 1227 (m), 1112 (w), 1047 (m), 924 (m), 745 (vs), 714 (w), 635 (s), 582 (w), 562 (w), 557 (w), 475 (w), 462 (w), 456 (w), 433 (w) cm^{-1}.

EA: Calc. for C$_{44}$H$_{34}$N$_2$O$_2$: C 84.86; H 5.50; N 4.50. Found: C 84.20; H 5.55; N 4.40%.

R$_f$ = 0.56 (Cyclohexane : ethyl acetate = 4:1).

3,6-Bis(3,6-di-*tert*-butyl-9*H*-carbazol-9-yl)phenanthrene-9,10-dione (28e)

A sealable vial was charged with 3',6'-bis(3,6-di-*tert*-butyl-9*H*-carbazol-9-yl)dispiro[[1,3]dioxolane-2,9'-phenanthrene-10',2''-[1,3]dioxolane] (0.45 g, 0.59 mmol, 1.00 eq.) and 3,4-diaminobenzenesulfonic acid (0.11 g, 0.59 mmol, 1.00 eq.). Subsequently, it was sealed and acetic acid (96%, 3.0 mL) was added. The reaction mixture was stirred at 120 °C for 19 h. Then, it was cooled to room temperature and purified by means of column chromatography (cyclohexane : dichloromethane = 2:1). Instead of 3,6-bis(3,6-di-*tert*-butyl-9*H*-carbazol-9-yl)dibenzo [*a*,*c*]phenazine-11-sulfonic acid, the product 28e (0.25 g, 0.33 mmol, 62%) was obtained as red crystals.

^{1}H NMR (400 MHz, CDCl$_3$) δ = 8.50 (d, ^3J = 8.3 Hz, 2H, 2-H), 8.22 (d, ^4J = 1.9 Hz, 2H, 5-H), 8.14–8.07 (m, 4H, 5'-H), 7.78 (dd, ^3J = 8.3 Hz, ^4J = 1.9 Hz, 2H, 3-H), 7.54–7.44 (m, 8H, 2'-H, 3'-H), 1.44 (s, 36H, C*H*$_3$) ppm.

^{13}C NMR (101 MHz, CDCl$_3$) δ = 179.12 (2C, C=O), 145.82 (2C, C-4), 144.65 (4C, C-4'), 138.10 (4C, C-6'), 137.30 (2C, C-6), 132.93 (2C, C-2), 128.94 (2C, C-1), 126.80 (2C, C-3), 124.54 (4C, C-1'), 124.41 (4C, C-3'), 120.78 (2C, C-5), 116.81 (4C, C-5'), 109.52 (4C, C-2'), 34.95 (4C, C$_q$), 32.04 (12C, *C*H$_3$) ppm.

MS (FAB, 3-NBA), m/z (%): 765 (100) [M+H]$^+$, 764 (86) [M]$^+$.

HRMS–FAB *(m/z)*: Calc. for [C$_{54}$H$_{56}$O$_2$N$_2$]$^+$: 764.4336; found: 764.4335.

IR (ATR, ṽ) = 2952 (m), 2861 (w), 1674 (m), 1594 (vs), 1499 (s), 1489 (s), 1469 (vs), 1451 (vs), 1361 (vs), 1320 (vs), 1292 (vs), 1262 (vs), 1228 (vs), 1201 (s), 1183 (m), 1129 (m), 1033 (m), 924 (s), 877 (s), 807 (vs), 739 (m), 612 (vs), 421 (s) cm^{-1}.

R$_f$ = 0.32 (Cyclohexane : dichloromethane = 1:2).

10-Fluorodibenzo[*a,c*]phenazine (29a)

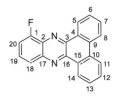 A sealable vial was charged with 3-fluorobenzene-1,2-diamine (1.00 g, 7.93 mmol, 2.97 equiv.) and phenanthrene-9,10-dione (1.65 g, 7.93 mmol, 1.00 equiv.). Then, acetic acid (25.0 mL) was added. The reaction mixture was refluxed at 100°C for 19 h. Then, the solid was filtered off, washed with water and dried under reduced pressure. The crude mixture was further purified employing column chromatography (cyclohexane : ethyl acetate = 100:1). The product (644 mg, 2.16 mmol, 27% yield) was obtained as a fluffy yellow solid.

^{19}**F NMR** (376 MHz, C_6D_6) δ = -124.31 ppm.

^{1}**H NMR** (400 MHz, CDCl$_3$) δ = 9.61 – 9.66 (m, 1H, 11-H), 9.51 – 9.56 (m, 1H, 8-H), 8.14 (ddd, $^3J_{H,H}$ = 7.7 Hz, $^4J_{H,H}$ = 5.7 Hz, $^5J_{H,H}$ = 2.2 Hz, 2H, 5-H, 14-H), 7.99 (dt, $^3J_{H,H}$ = 8.6 Hz, $^5J_{H,F}$ = 1.1 Hz, 1H, 18-H), 7.37 – 7.48 (m, 4H, 6-H, 7-H, 12-H, 13-H), 7.11 (ddd, $^3J_{H,H}$ = 8.6 Hz, $^3J_{H,H}$ = 7.7 Hz, $^4J_{H,F}$ = 5.5 Hz, 1H, 19-H), 7.01 (ddd, $^3J_{H,F}$ = 9.8 Hz, $^3J_{H,H}$ = 7.7 Hz, $^4J_{H,H}$ = 1.3 Hz, 1H, 20-H) ppm.

^{13}**C NMR** (101 MHz, C_6D_6) δ = 143.43, 143.40, 133.63, 132.81, 132.51, 130.75, 130.70, 130.67, 130.61, 128.84, 128.75, 128.18, 127.94, 127.27, 126.86, 125.43, 125.38, 123.23, 123.08, 113.24, 113.06 ppm.

MS (FAB, 3-NBA), m/z (%): 299 (47) [M+H]$^+$.

HRMS–FAB *(m/z)*: Calc. for [$C_{20}H_{12}N_2F$]$^+$: 299.0979; found: 299.0979.

IR (ATR, ṽ) = 3058 (w), 2921 (w), 2854 (w), 1625 (w), 1605 (w), 1494 (w), 1449 (m), 1347 (s), 1248 (m), 1128 (m), 1088 (s), 1065 (m), 1045 (m), 754 (vs), 720 (vs), 707 (m), 694 (m), 548 (s), 540 (m), 506 (m), 436 (s) cm^{-1}.

R$_f$ = 0.72 (Cyclohexane : ethyl acetate = 4:1).

11-Bromodibenzo[a,c]phenazine (29b)

The synthesis was carried out according to a literature procedure.[118] A sealable vial was charged with phenanthrene-9,10-dione (500 mg, 2.40 mmol, 1.00 equiv.) and 4-bromobenzene-1,2-diamine (449 mg, 2.40 mmol, 1.00 equiv.). Then, acetic acid (15.0 mL) was added, and the reaction mixture was stirred at 120 °C for 15 h. The product mixture is then filtered and washed with water and ethanol. The product (832 mg, 2.32 mmol, 96% yield) was obtained as a yellow solid.

^{1}H NMR (400 MHz, CDCl$_3$) δ = 9.39–9.32 (m, 2H, 7-H, 17-H), 8.56 (d, ^3J = 8.1 Hz, 2H, 10-H, 13-H), 8.51 (d, ^4J = 2.2 Hz, 1H, 20-H), 8.18 (d, ^3J = 8.9 Hz, 1H, 3-H), 7.91 (dd, ^3J = 8.9 Hz, ^4J = 2.2 Hz, 1H, 2-H), 7.81 (t, ^3J = 7.5 Hz, 2H, 9-H, 13-H), 7.75 (t, ^3J = 7.5 Hz, 2H, 8-H, 15-H) ppm.

R$_f$ = 0.67 (Cyclohexane : ethyl acetate = 4:1).

10-(Dibenzo[a,c]phenazin-10-yl)-10H-phenoxazine (30a)

A sealable vial was charged with 10-fluorodibenzo[a,c]phen-azine (300 mg, 1.01 mmol, 1.00 equiv.), 10H-phenoxazine (184 mg, 1.01 mmol, 1.00 equiv.) and NaH (60% dispersion in mineral oil, 80.4 mg, 2.01 mmol, 2.00 equiv.) and subsequently evacuated and backfilled with argon. Then, N,N-dimethyl-formamide (5.25 mL, abs.) was added. The reaction mixture was stirred at 100°C for 5 d. After cooling to room temperature, the reaction mixture was poured into an excess of water and extracted with dichloromethane (3×). The combined organic fractions were then washed with brine, dried over Na$_2$SO$_4$ and concentrated under reduced pressure. The crude mixture was further purified employing column chromatography (cyclohexane : dichloromethane = 5:1 → 4:1). The product (180 mg, 390 μmol, 39% yield) was obtained as a red solid.

^{1}H NMR (500 MHz, CDCl$_3$) δ = 9.39 (dd, ^3J = 7.9 Hz, ^4J = 1.6 Hz, 1H, 11-H), 9.06 (dd, ^3J = 8.2 Hz, ^4J = 1.5 Hz, 1H, 8-H), 8.53 (dd, ^3J = 8.1 Hz, 1.3, 1H, 14-H), 8.49–8.46 (m, 2H, 5-H,

18-H), 8.04–7.94 (m, 2H, 19-H, 20-H), 7.80 (ddd, ^3J = 8.1 Hz, ^3J = 7.1 Hz, ^4J = 1.6 Hz, 1H, 13-H), 7.78 – 7.69 (m, 2H, 6-H, 12-H), 7.60 (ddd, ^3J = 8.2 Hz, ^3J = 7.1 Hz, ^4J = 1.1 Hz, 1H, 7-H), 6.83 (dd, ^3J = 8.0 Hz, ^4J = 1.5 Hz, 2H, 2'-H), 6.63 (td, ^3J = 7.7 Hz, ^4J = 1.5 Hz, 2H, 3'-H), 6.46 (td, ^3J = 7.7 Hz, ^4J = 1.5 Hz, 2H, 4'-H), 5.86 (dd, ^3J = 8.0 Hz, ^4J = 1.5 Hz, 2H, 5'-H) ppm.

^{13}C NMR (126 MHz, CDCl3) δ = 144.34 (C-1), 144.28 (2C, C-1'), 143.02 (C-16), 142.67 (C-3), 139.28 (C-17), 136.99 (C-2), 135.23 (2C, C-6'), 133.45 (C-20), 132.51 (C-9), 132.29 (C-10), 130.82 (2C, C-6, C-13), 130.74 (C-19), 130.33 (C-18), 130.04 (C-15), 129.96 (C-4), 128.35 (C-7), 128.13 (C-12), 127.06 (C-8), 126.57 (C-11), 123.26 (2C, C-4'), 123.15 (C-14), 122.78 (C-5), 121.50 (2C, C-3'), 115.59 (2C, C-2'), 114.23 (2C, C-5') ppm.

MS (FAB, 3-NBA), m/z (%): 461 (12) [M]$^+$.

HRMS–FAB (m/z): Calc. for [C$_{32}$H$_{19}$ON$_3$]$^+$: 461.1523, found: 461.1522.

IR (ATR, ṽ) = 3065 (w), 3054 (w), 1587 (w), 1485 (vs), 1462 (s), 1448 (s), 1391 (w), 1340 (vs), 1315 (m), 1293 (s), 1273 (vs), 1268 (vs), 1224 (m), 1211 (m), 1201 (m), 1166 (m), 1146 (m), 1125 (m), 1102 (m), 1084 (m), 1068 (m), 1040 (m), 1027 (m), 980 (w), 956 (w), 919 (w), 909 (w), 864 (w), 841 (w), 762 (s), 752 (s), 734 (vs), 720 (vs), 683 (m), 650 (m), 643 (s), 629 (w), 616 (m), 605 (w), 589 (w), 579 (m), 552 (s), 530 (m), 506 (w), 486 (w), 472 (w), 450 (m), 435 (s), 418 (m), 404 (m), 381 (w) cm^{-1}.

R_f = 0.63 (Cyclohexane : ethyl acetate = 4:1).

10-(9,9-Dimethylacridin-10(9H)-yl)dibenzo[a,c]phenazine (30b)

A sealable vial was charged with 10-fluorodibenzo[a,c]phen-azine (200 mg, 670 µmol, 1.00 equiv.), 9,9-dimethyl-10H-acridine (140 mg, 670 µmol, 1.00 equiv.) and NaH (60% dispersion in mineral oil, 53.6 mg, 1.34 mmol, 2.00 equiv.) was evacuated and backfilled with argon. Subsequently, N,N-dimethylformamide (3.50 mL, abs.) was added. The reaction mixture was stirred at 100°C for 17 h. After cooling to room temperature, the reaction mixture was poured into an excess of water and extracted with dichloromethane (3×). The combined organic fractions were then washed with brine, dried over Na_2SO_4 and concentrated under reduced pressure. The crude mixture was further purified employing column chromatography (cyclohexane : ethyl acetate = 50:1). The product (97.0 mg, 199 μmol, 30% yield) was obtained as an orange solid.

^{1}H NMR (400 MHz, CDCl$_3$) δ = 9.44–9.40 (m, 1H, 8-H), 8.81 (dd, ^3J = 8.2 Hz, 1.4, 1H, 17-H), 8.54–8.50 (m, 2H, 4-H, 11-H), 8.45 (d, ^3J = 8.2 Hz, 1H, 14-H), 8.06 (dd, ^3J = 8.6 Hz, ^3J = 7.3 Hz, 1H, 3-H), 7.96 (dd, ^3J = 7.3 Hz, ^4J = 1.5 Hz, 1H, 2-H), 7.82–7.73 (m, 2H, 9-H, 10-H), 7.66 (ddd, ^3J = 8.2 Hz, ^3J = 7.1 Hz, ^4J = 1.4 Hz, 1H, 15-H), 7.56 (dd, ^3J = 7.4 Hz, ^4J = 1.6 Hz, 2H, 5'-H), 7.45–7.40 (m, 1H, 16-H), 6.90 (td, ^3J = 7.4 Hz, ^4J = 1.3 Hz, 2H, 4'-H), 6.82 (ddd, ^3J = 8.5 Hz, ^3J = 7.4 Hz, ^4J = 1.6 Hz, 2H, 3'-H), 6.17 (dd, ^3J = 8.5 Hz, ^4J = 1.3 Hz, 2H, 2'-H), 1.58 (s, 6H, CH_3) ppm.

^{13}C NMR (126 MHz, CDCl$_3$) δ = 144.01, 142.90, 141.50, 140.07, 138.81, 133.74, 132.48, 132.21, 130.82, 130.59, 130.47, 130.30, 130.18, 130.15, 130.01, 128.15, 127.94, 127.02, 126.64, 126.45, 125.54, 123.12, 122.75, 120.58, 114.31, 36.36 ppm.

MS (FAB, 3-NBA), m/z (%): 488 (7) [M+H]$^+$, 487 (6) [M]$^+$.

HRMS–FAB *(m/z)*: Calc. for [C$_{35}$H$_{26}$N$_3$]$^+$: 488.2121; found: 488.2123.

IR (ATR, ṽ) = 3071 (w), 3036 (w), 2963 (w), 2918 (w), 2902 (w), 1604 (w), 1592 (m), 1500 (w), 1473 (s), 1446 (vs), 1391 (w), 1353 (s), 1330 (vs), 1273 (vs), 1242 (w), 1221 (w), 1167 (w), 1126 (w), 1109 (m), 1077 (w), 1045 (m), 1028 (w), 761 (vs), 745 (m), 732 (vs), 725 (vs), 713 (vs), 697 (w), 686 (w), 673 (w), 653 (w), 635 (w), 626 (w), 616 (w), 589 (w), 560 (w), 550 (m), 543 (w), 535 (w), 507 (w), 477 (w), 438 (m), 426 (w) cm^{-1}.

R$_f$ = 0.69 (Cyclohexane : ethyl acetate = 4:1).

10-(3,6-Di-*tert*-butyl-9*H*-carbazol-9-yl)dibenzo[*a,c*]phenazine (30c)

A sealable vial containing 10-fluorodibenzo[*a,c*]phenazine (200 mg, 670 µmol, 1.00 equiv.), 3,6-ditert-butyl-9*H*-carbazole (187 mg, 670 µmol, 1.00 equiv.) and K_3PO_4 (356 mg, 1.68 mmol, 2.50 equiv.), was evacuated and backfilled with argon. Subsequently, DMSO (3.50 mL, abs.) was added. The reaction mixture was then stirred at 100 °C for 26 h. After cooling to room temperature, the reaction mixture was poured into an excess of water and extracted with dichloromethane (3×). The combined organic fractions were then washed with brine, dried over Na_2SO_4, and concentrated under reduced pressure. The crude mixture was further purified by column chromatography (cyclohexane : ethyl acetate = 100:1 → 50:1). The product (253 mg, 454 µmol, 68% yield) was obtained as a yellow solid.

¹H NMR (400 MHz, CDCl3) δ = 9.46–9.42 (m, 1H, 11-H), 8.56–8.51 (m, 1H, 14-H), 8.48–8.43 (m, 2H, 8-H, 18-H), 8.26 (dd, ^4J = 2.0 Hz, ^5J = 0.6 Hz, 2H, 5'-H), 8.24 (dd, ^3J = 8.1 Hz, ^4J = 1.4 Hz, 1H, 5-H), 8.10 (dd, ^3J = 7.3 Hz, ^4J = 1.5 Hz, 1H, 20-H), 8.04 (dd, ^3J = 8.4 Hz, ^3J = 7.3 Hz, 1H, 19-H), 7.85–7.72 (m, 2H, 12-H, 13-H), 7.64 (ddd, ^3J = 8.3 Hz, ^3J = 7.1 Hz, ^4J = 1.4 Hz, 1H, 7-H), 7.38 (dd, ^3J = 8.7 Hz, ^4J = 2.0 Hz, 2H, 3'-H), 7.31–7.27 (m, 1H, 6-H), 7.16 (dd, ^3J = 8.7 Hz, ^5J = 0.6 Hz, 2H, 2'-H), 1.49 (s, 18H, C*H*3) ppm.

MS (FAB, 3 NBA), m/z (%): 558 (18) [M+H]⁺, 557 (19) [M]⁺

HRMS–FAB *(m/z)*: calc. for $[C_{40}H_{35}N_3]^+$: 557.2831; found: 557.2833.

IR (ATR, ṽ) = 3061 (w), 2951 (s), 2900 (w), 1741 (vw), 1604 (w), 1496 (m), 1490 (m), 1475 (s), 1449 (m), 1358 (vs), 1293 (s), 1262 (s), 1235 (s), 875 (w), 816 (s), 755 (vs), 721 (vs), 615 (m) cm⁻¹.

R*f* = 0.81 (Cyclohexane : ethyl acetate = 4:1).

11-(3,6-Di-*tert*-butyl-9*H*-carbazol-9-yl)dibenzo[*a,c*]phenazine (32)

A sealable vial was charged with 11-bromo-phenanthro[9,10-*b*]quinoxaline (200 mg, 557 µmol, 1.00 equiv.), 3,6-di-*tert*-butyl-9*H*-carbazole (156 mg, 557 µmol, 1.00 equiv.), NaO*t*Bu (107 mg, 1.11 mmol, 2.00 equiv.) and Pd(OAc)$_2$ (12.5 mg, 55.7 µmol, 0.10 equiv.). It was then sealed, evacuated and backfilled with argon three times. Subsequently, toluene (5.00 mL, abs.) was added and P(*t*Bu)$_3$ (22.5 mg, 111 µL, 111 µmol, 1.00 M in toluene, 0.20 equiv.) was added dropwise while stirring. The reaction mixture was stirred at 120 °C for 64 h. The crude mixture was pulled over a Celite® plug with dichloromethane. Then, all volatile components were removed under reduced pressure and the crude product was further purified employing column chromatography (cyclohexane : ethyl acetate = 50:1 → 5:1). The product (309 mg, 554 µmol, 100% yield) was obtained as a bright yellow solid.

¹H NMR (400 MHz, CDCl$_3$) δ = 9.45 (dd, ^3J = 7.8 Hz, ^4J = 1.7 Hz, 1H, 12-H), 9.42 (dt, ^3J = 8.0 Hz, ^4J = 1.5 Hz, 1H, 9-H), 8.60 (d, ^3J = 8.0 Hz, 2H, 6-H, 15-H), 8.55–8.48 (m, 2H, 2-H, 19-H), 8.20 (d, ^4J = 1.8 Hz, 2H, 5'-H), 8.12 (dt, ^3J = 9.0 Hz, ^4J = 1.8 Hz, 1H, 20-H), 7.88–7.80 (m, 2H, 7-H, 14-H), 7.79–7.73 (m, 2H, 8-H, 13-H), 7.66–7.59 (m, 2H, 2'-H), 7.54 (dd, ^3J = 8.7 Hz, ^4J = 1.8 Hz, 2H, 3'-H), 1.50 (s, 18H, C*H*$_3$) ppm.

¹³C NMR (101 MHz, CDCl$_3$) δ = 143.87 (2C, C-4'), 143.22 (C-17), 143.02 (C-1), 142.64 (C-4), 141.04 (C-18), 139.55 (C-3), 139.04 (2C, C-1'), 132.42 (C-16), 132.23 (C-5), 131.07 (C-19), 130.72 (C-14), 130.56 (C-7), 130.44 (C-11), 130.32 (C-10), 128.94 (C-20), 128.19 (2C, C-8, C-13), 126.61 (C-9), 126.43 (C-12), 125.06 (C-2), 124.13 (2C, C-6'), 124.10 (2C, C-3'), 123.15 (2C, C-6, C-15), 116.59 (2C, C-5'), 109.58 (2C, C-2'), 34.97 (2C, C$_q$), 32.16 (6C, C*H*$_3$) ppm.

MS (FAB, 3-NBA), m/z (%): 558 (24) [M+H]$^+$, 557 (24) [M]$^+$.

HRMS–FAB *(m/z)*: Calc. for [C$_{40}$H$_{35}$N$_3$]$^+$: 557.2825; found: 557.2825.

IR (ATR, ṽ) = 3060 (w), 2952 (vs), 2921 (vs), 2854 (s), 1615 (m), 1581 (w), 1499 (vs), 1490 (s), 1476 (vs), 1449 (vs), 1357 (vs), 1312 (vs), 1299 (vs), 1293 (s), 1259 (s), 1235 (s), 1200 (s),

1176 (m), 1037 (s), 877 (s), 841 (m), 813 (vs), 807 (vs), 764 (vs), 724 (vs), 611 (vs), 554 (vs), 439 (s), 424 (m) cm^{-1}.

R_f = 0.73 (Cyclohexane : ethyl acetate = 4:1).

Dimethyl 4,4'-(11-bromodibenzo[*a,c*]phenazine-2,7-diyl)bis(3,5-dimethylbenzoate) (34)

A flask was charged with dimethyl 4,4'-(9,10-dioxo-9,10-dihydrophenanthrene-2,7-diyl) bis(3,5-dimethylbenzoate) (309 mg, 580 μmol, 1.00 equiv.) and 4-bromobenzene-1,2-diamine (109 mg, 580 μmol, 1.00 equiv.). Then, acetic acid (10.0 mL) was added and the reaction mixture refluxed at 120 °C for 4 h. Subsequently, the product mixture was filtered off, and the residue was washed with water. It was then dried under reduced pressure. The product (249 mg, 364 μmol, 63% yield) was obtained as a pale-yellow solid.

¹H NMR (400 MHz, CDCl$_3$) δ = 9.19 (d, ^4J = 1.9 Hz, 2H, 7-H, 16-H), 8.70 (d, ^3J = 8.3 Hz, 2H, 10-H, 13-H), 8.45 (d, ^4J = 2.1 Hz, 1H, 20-H), 8.12 (d, ^3J = 9.1 Hz, 1H, 3-H), 7.92–7.84 (m, 5H, 2-H, 3'-H), 7.62 (dd, ^3J = 8.3 Hz, ^4J = 1.9 Hz, 2H, 9-H, 14-H), 3.98 (s, 6H, OCH_3), 2.19 (s, 12H, CH_3) ppm.

¹³C NMR (101 MHz, CDCl$_3$) δ = 167.45 (2C, COOCH$_3$), 146.09 (C-1'), 146.04 (C-1'), 143.11 (C-5), 142.77 (C-18), 142.73 (C-19), 141.02 (C-4), 140.42 (2C, C-8, C-15), 136.74 (2C, C-2'), 136.71 (2C, C-2'), 133.65 (C-2), 131.75 (C-20), 131.39 (C-9), 131.24 (C-14), 131.19 (C-11), 131.04 (C-12), 130.84 (C-3), 130.62 (C-6), 130.53 (C-17), 129.21 (2C, C-4'), 128.76 (4C, C-3'), 126.55 (C-7), 126.41 (C-16), 124.20 (C-1), 123.66 (C-10), 123.64 (C-13), 52.27 (2C, OCH_3), 21.17 (4C, CH$_3$) ppm.

MS (FAB, 3-NBA), m/z (%): 713 (3) [M(^{81}Br)+H]$^+$, 712 (11) [M(^{81}Br)]$^+$, 711 (10) [M(^{79}Br)+H]$^+$, 710 (5) [M(^{79}Br)]$^+$.

158

IR (ATR, ṽ) = 2990 (w), 2946 (w), 2918 (w), 1714 (vs), 1601 (w), 1431 (s), 1340 (w), 1312 (vs), 1222 (vs), 1211 (vs), 1200 (vs), 1120 (s), 1020 (s), 916 (m), 901 (s), 836 (s), 827 (s), 803 (m), 768 (vs), 735 (m), 575 (w), 476 (s) cm^{-1}.

EA: Calc. for $C_{40}H_{31}BrN_2O_4$: C 70.28; H 4.57; N 4.10. Found: C 69.61; H 4.55; N 4.09%.

R$_f$ = 0.47 (Cyclohexane : ethyl acetate = 4:1).

Dimethyl   4,4'-(11-(3,6-di-*tert*-butyl-9*H*-carbazol-9-yl)dibenzo[*a,c*]phenazine-2,7-diyl)bis(3,5-dimethylbenzoate) (35a)

A sealable vial was charged with dimethyl 4,4'-(11-bromodibenzo[*a,c*]phenazine-2,7-diyl) bis(3,5-dimethylbenzoate) (45.0 mg, 65.8 µmol, 1.00 equiv.), 3,6-di-*tert*-butyl-9*H*-carbazole (18.4 mg, 65.8 µmol, 1.00 equiv.), Pd(OAc)₂ (739 µg, 3.29 µmol, 0.0500 equiv.) and NaO*t*Bu (25.3 mg, 263 µmol, 4.00 equiv.). It was then sealed and evacuated and backfilled with argon three times. Subsequently, P(*t*Bu)₃ (1.33 mg, 6.58 µL, 6.58 µmol, 1.00 M in toluene, 0.100 equiv.) was added dropwise, and the reaction mixture was stirred at 110 °C for 17 h. After cooling to room temperature, the product mixture was poured into an excess of water and extracted with dichloromethane three times. The combined organic phases were washed with brine and dried over Na₂SO₄ before being concentrated under reduced pressure. The crude product was purified employing flash chromatography using a cyclohexane/ethyl acetate gradient. The product (17.0 mg, 19.3 µmol, 29% yield) was obtained as a bright yellow solid.

¹H NMR (400 MHz, CDCl₃) δ = 9.27 (d, ^4J = 1.9 Hz, 1H, 7-H), 9.25 (d, ^4J = 1.9 Hz, 1H, 16-H), 8.74 (d, ^3J = 8.3 Hz, 2H, 10-H, 13-H), 8.48 (d, ^4J = 2.3 Hz, 1H, 20-H), 8.45 (d, ^3J = 9.1 Hz, 1H, 3-H), 8.17 (d, ^4J = 1.9 Hz, 2H, 5"-H), 8.12 (dd, ^3J = 9.0 Hz, ^4J = 2.4 Hz, 1H, 2-H), 7.91 (s, 2H, 3'-H), 7.87 (s, 2H, 3'-H), 7.67–7.62 (m, 2H, 9-H, 14-H), 7.60 (d, ^3J = 8.7 Hz, 2H, 2"-H),

7.50 (dd, ^3J = 8.7 Hz, ^4J = 1.9 Hz, 2H, 3''-H), 3.99 (s, 3H, OCH_3), 3.96 (s, 3H, OCH_3), 2.23 (s, 6H, CH_3), 2.21 (s, 6H, CH_3), 1.48 (s, 18H, CH_3) ppm.

^{13}C NMR (101 MHz, CDCl$_3$) δ = 167.49 (C=O), 167.47 (C=O), 146.23 (C-1'), 146.16 (C-1'), 143.93 (2C, C-4''), 143.16 (C-1), 143.09 (C-18), 142.52 (C-5), 141.05 (C-4), 140.37 (2C, C-8, C-15), 139.77 (C-19), 138.90 (2C, C-1''), 136.81 (2C, C-2'), 136.73 (2C, C-2'), 131.23 (C-14), 131.15 (C-9), 131.04 (C-3), 130.96 (C-12), 130.84 (C-11), 130.72 (2C, C-4'), 129.20 (C-17), 129.15 (C-6), 129.08 (C-2), 128.78 (4C, C-3'), 126.59 (C-16), 126.35 (C-7), 124.80 (C-20), 124.15 (2C, C-6''), 124.11 (2C, C-3''), 123.66 (C-13), 123.63 (C-10), 116.56 (2C, C-5''), 109.54 (2C, C-2''), 52.28 (OCH_3), 52.23 (OCH_3), 34.94 (2C, C$_q$), 32.12 (6C, CH_3), 21.22 (4C, CH_3).

R_f = 0.61 (Cyclohexane : ethyl acetate = 4:1).

3,6-dibromo-10-fluorodibenzo[a,c]phenazine (2Br-BP-F) (36)

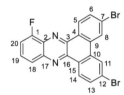

A sealable vial was charged with 3-fluorobenzene-1,2-diamine (800 mg, 6.34 mmol, 1.00 equiv.) and 3,6-dibromophenanthrene-9,10-dione (2.32 g, 6.34 mmol, 1.00 equiv.). Then, acetic acid (10.7 mL, 96%) and ethanol (2.67 mL) were added, and the reaction was stirred at 100 °C for 21 h. Subsequently, the precipitate was filtered off and dried under reduced pressure to give the product (2.65 g, 5.83 mmol, 92%) as a beige solid.

^{19}F NMR (376 MHz, C$_6$D$_6$) δ = -123.92 ppm.

^{1}H NMR (400 MHz, C$_6$D$_6$) δ = 9.22 (d, $^3J_{H,H}$ = 8.6 Hz, 1H, 14-H), 9.12 (d, $^3J_{H,H}$ = 8.5 Hz, 1H, 5-H), 8.15 (dd, J = 4.6 Hz, $^4J_{H,H}$ = 1.9 Hz, 2H, 8-H, 11-H), 7.94 (d, $^3J_{H,H}$ = 8.4 Hz, 1H, 18-H), 7.53 (ddd, $^3J_{H,H}$ = 8.2 Hz, J = 5.7 Hz, $^4J_{H,H}$ = 1.9 Hz, 2H, 6-H, 13-H), 7.11 (dd, $^3J_{H,H}$ = 8.1 Hz, $^4J_{H,F}$ = 5.5 Hz, 1H, 19-H), 7.04–6.97 (m, 1H, 20-H) ppm.

HRMS–EI *(m/z)*: Calc. for [C$_{20}$H$_9$N$_2$^{79}Br$_2$F]$^+$: 453.9111, found: 453.9110.

IR (ATR, ṽ) = 1931 (vw), 1628 (w), 1595 (m), 1567 (w), 1537 (m), 1493 (m), 1485 (m), 1395 (m), 1358 (m), 1347 (vs), 1329 (w), 1248 (m), 1140 (m), 1098 (vs), 1088 (s), 1064 (m), 1028

160

(m), 867 (m), 839 (s), 820 (vs), 786 (m), 754 (vs), 711 (m), 591 (m), 551 (vs), 537 (s), 467 (w) cm^{-1}.

EA: Calc. for $C_{20}H_9Br_2FN_2$: C 52.67; H 1.99; N 6.14. Found: C 52.55; H 1.80; N 6.14%.

M.p.: 305 °C.

3,6-Dibromo-10-(3,6-di-*tert*-butyl-9*H*-carbazol-9-yl)dibenzo[*a,c*]phenazine (37c)

A sealable vial was charged with 3,6-dibromo-10-fluorodibenzo[*a,c*]phenazine (327 mg, 717 µmol, 1.00 equiv.), 3,6-di-*tert*-butyl-9*H*-carbazole (202 mg, 723 µmol, 1.01 equiv.) and K$_3$PO$_4$ (304 mg, 1.43 mmol, 2.00 equiv.). It was then evacuated and backfilled with argon. Subsequently, DMSO (3.00 mL, abs.) was added. The reaction mixture was then stirred at 110 °C for 7 h. After cooling to room temperature, the mixture was poured into an excess of water and extracted with dichloromethane (3x). The combined organic fractions were washed with brine, dried over Na$_2$SO$_4$ and concentrated under reduced pressure. The crude mixture was further purified employing column chromatography (*n*-pentane : dichloromethane = 2:1). The product (405 mg, 566 µmol, 79% yield) was obtained as a yellow solid.

^{1}H NMR (400 MHz, CDCl$_3$) δ = 9.27 (d, ^3J = 8.6 Hz, 1H, 14-H), 8.55 (d, ^4J = 1.9 Hz, 1H, 11-H), 8.48 (d, ^4J = 1.9 Hz, 1H, 8-H), 8.43 (dd, ^3J = 8.4 Hz, ^4J = 1.5 Hz, 1H, 18-H), 8.26 (dd, ^4J = 2.0 Hz, ^5J = 0.6 Hz, 2H, 5'-H), 8.12 (dd, ^3J = 7.3 Hz, ^4J = 1.5 Hz, 1H, 20-H), 8.06 (dd, ^3J = 8.5 Hz, ^3J = 7.0 Hz, 2H, 5-H, 19-H), 7.87 (dd, ^3J = 8.6 Hz, ^4J = 1.9 Hz, 1H, 13-H), 7.42–7.34 (m, 3H, 3'-H, 6-H), 7.12 (dd, ^3J = 8.6 Hz, ^5J = 0.6 Hz, 2H, 2'-H), 1.49 (s, 18H, CH_3) ppm.

^{13}C NMR (101 MHz, CDCl$_3$) δ = 143.40 (C-1), 143.14 (2C, C-4'), 142.05 (C-16), 141.10 (C-3), 140.82 (2C, C-1'), 138.46 (C-2), 136.22 (C-17), 132.70 (C-10), 132.34 (C-9), 131.86 (C-13), 131.65 (C-6), 130.13 (C-19), 129.33 (C-18), 129.19 (2C, C-4, C-15), 128.99 (C-20), 128.72 (C-5), 128.28 (C-14), 126.17 (C-11), 125.98 (C-12), 125.82 (C-8), 125.77 (C-7), 123.86 (2C, C-6'), 123.55 (2C, C-3'), 116.16 (2C, C-5'), 110.47 (2C, C-2'), 34.94 (2C, C$_q$), 32.22. (2C, CH_3) ppm.

MS (FAB, 3-NBA), m/z (%): 716 (10) [M+H]$^+$, 715 (13) [M]$^+$.

HRMS–FAB *(m/z)*: Calc. for [C$_{40}$H$_{33}$N$_3$^{79}Br$_2$]$^+$: 713.1036; found 713.1035.

IR (ATR, ṽ) = 2952 (m), 2934 (w), 2901 (w), 2864 (w), 1595 (m), 1567 (w), 1537 (w), 1490 (m), 1473 (s), 1374 (w), 1349 (s), 1327 (m), 1295 (m), 1259 (m), 1235 (m), 1098 (m), 1069 (m), 1030 (w), 868 (m), 824 (m), 813 (vs), 803 (m), 759 (s), 613 (m), 551 (s), 422 (w) cm^{-1}.

R$_f$ = 0.81 (Cyclohexane : ethyl acetate = 4:1).

10,10'-(10-(3,6-di-*tert*-butyl-9*H*-carbazol-9-yl)dibenzo[*a,c*]phenazine-3,6-diyl)bis(10*H*-phenoxazine) (38)

A sealable vial was charged with 3,6-dibromo-10-(3,6-di-*tert*-butyl-9*H*-carbazol-9-yl) dibenzo[*a,c*] phenazine (100 mg, 140 µmol, 1.00 equiv.), 10*H*-phenoxazine (51.2 mg, 280 µmol, 2.00 equiv.), NaO*t*Bu (40.3 mg, 419 µmol, 3.00 equiv.) and Pd(dba)$_2$ (4.02 mg, 6.99 µmol, 0.05 equiv.). It was then sealed, evacuated and backfilled with argon. Then, toluene (2.00 mL, abs.) was added and P(*t*Bu)$_3$ (4.24 mg, 5.09 µL, 21.0 µmol, 0.15 equiv.) was added dropwise while stirring. The reaction mixture was stirred at 120 °C for 6 h. After cooling to room temperature, the mixture was concentrated under reduced pressure and the crude product was further purified *via* a filter column using dichloromethane. The product (121 mg, 132 µmol, 94% yield) was obtained as a dark red solid.

^{1}H NMR (400 MHz, CDCl$_3$) δ = 9.70 (d, ^3J = 8.5 Hz, 1H, 8-H), 8.52 (dd, ^3J = 8.4 Hz, ^4J = 1.5 Hz, 1H, 4-H), 8.44–8.39 (m, 2H, 11-H, 17-H), 8.36 (d, ^4J = 2.0 Hz, 1H, 14-H), 8.26 (d, ^4J = 1.8 Hz, 2H, 5'-H), 8.19 (dd, ^3J = 7.3 Hz, ^4J = 1.5 Hz, 1H, 2-H), 8.13 (dd, ^3J = 8.4 Hz, ^3J = 7.3 Hz, 1H, 3-H), 7.79 (dd, ^3J = 8.5 Hz, ^4J = 1.9 Hz, 1H, 9-H), 7.40 (dd, ^3J = 8.7 Hz, ^4J = 1.8 Hz, 2H, 3'-H), 7.28 (dd, ^3J = 8.5 Hz, ^4J = 2.0 Hz, 1H, 16-H), 7.17 (d, ^3J = 8.7 Hz, 2H, 2'-H), 6.78–6.46 (m, 12H, 2''-H, 3''-H, 4''-H), 6.02 (dd, ^3J = 8.0 Hz, ^4J = 1.5 Hz, 2H, 5''-H), 5.90 (dd, ^3J = 8.0 Hz, ^4J = 1.4 Hz, 2H, 5''-H), 1.48 (s, 18H, CH_3) ppm.

¹³C NMR (101 MHz, CDCl₃) δ = 144.11 (2C, C-1''), 144.07 (2C, C-1'), 143.54 (C-1), 143.19 (2C, C-4'), 141.66 (C-10), 141.41 (C-15), 141.31 (2C, C-12, C-13), 140.88 (2C, C-6, C-19), 138.53 (C-20), 136.32 (C-5), 134.45 (C-12), 134.17 (2C, C-6''), 134.08 (2C, C-6''), 131.19 (C-9), 131.00 (C-16), 130.35 (C-7), 130.23 (C-18), 130.19 (C-3), 130.12 (C-17), 129.88 (C-8), 129.28 (C-4), 129.10 (C-2), 125.99 (C-11), 125.38 (C-14), 123.88 (2C, C-6'), 123.57 (2C, C-3'), 123.46 (2C, C-4''), 123.39 (2C, C-4''), 121.90 (2C, C-3''), 121.80 (2C, C-3''), 116.21 (2C, C-5'), 115.84 (2C, C-2'), 115.76 (2C, C-2''), 113.41 (2C, C-5''), 113.31 (2C, C-5''), 110.49 (2C, C-2'), 34.94 (2C, C_q), 32.21 (6C, CH₃) ppm.

MS (ESI), m/z (%): 920 (72) [M]⁺.

IR (ATR, ṽ) = 3050 (vw), 2952 (w), 2901 (w), 2864 (w), 1602 (w), 1483 (vs), 1463 (s), 1357 (m), 1323 (s), 1290 (s), 1268 (vs), 1234 (s), 1200 (s), 1033 (m), 806 (m), 737 (vs), 630 (s), 619 (s), 612 (s) cm⁻¹.

R_f = 0.83 (Cyclohexane : ethyl acetate = 4:1).

10,10'-(10-fluorodibenzo[a,c]phenazine-3,6-diyl)bis(10H-phenoxazine) (39a)

A sealable vial was charged with 3,6-dibromo-10-fluorodibenzo[a,c]phenazine (651 mg, 1.43 mmol, 1.00 equiv.), 10H-phenoxazine (575 mg, 3.14 mmol, 2.20 equiv.), TTBP · HBF₄ (50.1 mg, 173 μmol, 0.121 equiv.), NaOtBu (412 mg, 4.28 mmol, 3.00 equiv.) and Pd(dba)₂ (47.0 mg, 81.7 μmol, 0.06 equiv.). It was then sealed and evacuated, and backfilled with argon. Then, toluene (15.2 mL, abs.) was added. The reaction mixture was stirred at 120 °C for 14 h. The product mixture was pulled over a Celite® plug with dichloromethane and the filtrate was concentrated under reduced pressure. The crude product was purified employing column chromatography (cyclohexane : dichloromethane = 4:1). The product (943 mg, 1.43 mmol, quant.) was obtained as a bright red solid.

¹⁹F NMR (376 MHz, CDCl₃) δ = -124.28 ppm.

^{1}H NMR (500 MHz, CDCl$_3$) δ = 9.72 (d, $^3J_{H,H}$ = 8.5 Hz, 1H, 8-H), 9.67 (d, $^3J_{H,H}$ = 8.5 Hz, 1H, 17-H), 8.49 (dd, J = 4.0 Hz, $^4J_{H,H}$ = 1.9 Hz, 2H, 11-H, 14-H), 8.21 (dt, $^3J_{H,H}$ = 8.6 Hz, $^4J_{H,H}$ = 1.0 Hz, 1H, 4-H), 7.87 (ddd, $^3J_{H,H}$ = 8.6 Hz, $^3J_{H,H}$ = 7.7 Hz, $^4J_{H,F}$ = 5.4 Hz, 1H, 3-H), 7.79 (dt, $^3J_{H,H}$ = 8.4 Hz, $^4J_{H,H}$ = 2.2 Hz, 2H, 9-H, 16-H), 7.61 (ddd, $^3J_{H,F}$ = 9.1 Hz, $^3J_{H,H}$ = 7.7 Hz, $^4J_{H,H}$ = 1.2 Hz, 1H, 2-H), 6.73 (dt, $^3J_{H,H}$ = 7.8 Hz, $^4J_{H,H}$ = 1.2 Hz, 4H, 2'-H), 6.70–6.64 (m, 4H, 3'-H), 6.58 (td, $^3J_{H,H}$ = 7.7 Hz, $^4J_{H,H}$ = 1.5 Hz, 4H, 4'-H), 6.03 (dt, $^3J_{H,H}$ = 8.1 Hz, $^4J_{H,H}$ = 1.5 Hz, 4H, 5'-H) ppm.

^{13}C NMR (126 MHz, CDCl$_3$) δ = 157.63 (d, $^1J_{C,F}$ = 262.2 Hz, C-1), 144.13 (4C, C-6'), 143.39 (C-6), 142.62 (C-19), 141.87 (C-10), 141.83 (d, $^3J_{C,F}$ = 2.4 Hz. C-5), 141.79 (C-15), 134.47 (C-17), 134.27 (C-13), 134.17 (2C, C-1'), 134.16 (2C, C-1'), 133.57 (d, $^2J_{C,F}$ = 12.5 Hz, C-20), 131.42 (C-9), 131.32 (C-16), 130.23 (C-7), 130.21 (C-8), 130.13 (C-18), 129.96 (C-17), 129.79 (d, $^3J_{C,F}$ = 8.3 Hz, C-3), 126.04 (C-11), 125.93 (C-14), 125.56 (d, $^4J_{C,F}$ = 4.6 Hz, C-4), 123.48 (4C, C-3'), 121.94 (2C, C-4'), 121.92 (2C, C-4'), 115.86 (4C, C-5'), 114.03 (d, $^2J_{C,F}$ = 18.2 Hz, C-2), 113.44 (2C, C-22), 113.43 (2C, C-22) ppm.

MS (ESI), m/z (%): 663 (100), 662 (65), 647 (56), 154 (43).

HRMS–FAB (*m/z*): Calc. for [C$_{44}$H$_{25}$O$_2$N$_4$F]$^+$: 660.1956, found: 660.1956.

IR (ATR, ṽ) = 3060 (w), 1625 (w), 1602 (m), 1592 (w), 1506 (w), 1485 (vs), 1463 (m), 1402 (w), 1354 (s), 1332 (vs), 1292 (m), 1272 (vs), 1252 (s), 1224 (w), 1200 (m), 1128 (w), 1091 (m), 1043 (m), 936 (w), 866 (w), 752 (m), 739 (vs), 728 (vs), 632 (s), 446 (w) cm^{-1}.

EA: Calc. for (C$_{44}$H$_{25}$FN$_4$O$_2$): C 79.99; H 3.81; N 8.48. Found: C 80.15; H 3.75; N 8.35%.

R$_f$ = 0.76 (Cyclohexane : ethyl acetate = 4:1).

M.p.: 315 °C.

3,6-bis(3,6-di-*tert*-butyl-9*H*-carbazol-9-yl)-10-fluorodibenzo[*a,c*]phenazine (39c)

To a sealable vial, 3,6-dibromo-10-fluorodibenzo[*a,c*] phenazine (1.00 g, 2.19 mmol, 1.00 equiv.), 3,6-di*tert*-butyl-9*H*-carbazole (1.23 g, 4.38 mmol, 2.00 equiv.), NaO*t*Bu (632 mg, 6.58 mmol, 3.00 equiv.), Pd(OAc)$_2$ (49.2 mg, 219 μmol, 0.10 equiv.) were added, and the vial was evacuated and backfilled with argon. Then, toluene (10.0 mL, abs.) was added, and P(*t*Bu)$_3$ (88.7 mg, 438 μL, 438 μmol, 1.00 M in toluene, 0.50 equiv.) was added dropwise. Subsequently, the solution was stirred at 110 °C for 18 h. Then, the product mixture was cooled to room temperature and pulled over a Celite® plug with dichloromethane. The filtrate was concentrated under reduced pressure. The crude product was purified *via* column chromatography (cyclohexane : dichloromethane = 5:1 → 2:1). The product (1.81 g, 2.12 mmol, 97%) was obtained as a bright orange-yellow solid.

^{19}F NMR (376 MHz, CDCl$_3$) δ = -124.46 ppm.

^{1}H NMR (400 MHz, CDCl$_3$) δ = 9.60 (d, $^{3}J_{H,H}$ = 8.6 Hz, 1H, 8-H), 9.54 (d, $^{3}J_{H,H}$ = 8.6 Hz, 1H, 17-H), 8.66 (dd, J = 3.1, $^{4}J_{H,H}$ = 2.0 Hz, 2H, 11-H, 14-H), 8.17 (d, $^{4}J_{H,H}$ = 1.9 Hz, 4H, 5'-H), 8.13 (dd, $^{3}J_{H,H}$ = 8.6 Hz, $^{4}J_{H,H}$ = 1.2 Hz, 1H, 4-H), 8.01 (ddd, $^{3}J_{H,H}$ = 8.6 Hz, $^{4}J_{H,H}$ = 3.2 Hz, $^{4}J_{H,H}$ = 1.9 Hz, 2H, 9-H, 16-H), 7.81 (ddd, $^{3}J_{H,H}$ = 8.6 Hz, $^{3}J_{H,H}$ = 7.7 Hz, $^{4}J_{H,F}$ = 5.4 Hz, 1H, 3-H), 7.59–7.51 (m, 5H, 2-H, 2'-H), 7.43 (ddd, $^{3}J_{H,H}$ = 8.7 Hz, $^{4}J_{H,H}$ = 3.6 Hz, $^{4}J_{H,H}$ = 1.9 Hz, 4H, 3'-H), 1.47 (s, 18H, CH$_3$), 1.46 (s, 18H, CH$_3$) ppm.

^{13}C NMR (101 MHz, CDCl3) δ = 157.59 (d, $^{1}J_{C,F}$ = 261.8 Hz), 143.70 (2C, C-4'), 143.67 (2C, C-4'), 143.18 (C-10), 142.57 (C-15), 141.80 (d, $^{3}J_{C,F}$ = 2.12 Hz, C-5), 140.96 (C-12), 140.92 (C-13), 138.87 (4C, C-1'), 133.36 (C-6), 133.28 (d, $^{2}J_{C,F}$ = 12.2 Hz, C-20), 133.15 (C-19), 129.26 (d, $^{3}J_{C,F}$ = 8.3 Hz, C-3), 128.84 (C-8), 128.58 (C-7), 128.55 (C-18), 128.46 (C-17), 126.67 (C-9), 126.56 (C-16), 125.40 (d, $^{4}J_{C,F}$ = 4.4 Hz, C-4), 124.17 (4C, 2'), 123.97 (4C, C-6'), 120.27 (C-11), 120.19 (C-14), 116.57 (4C, C-5'), 113.62 (d, $^{2}J_{C,F}$ = 18.2 Hz, C-2), 109.27 (4C, C-2'), 34.90 (4C, C$_q$), 32.11 (12C, CH$_3$) ppm.

MS (FAB, 3-NBA), m/z (%): 854 (100) [M+H]$^+$, 853 (75) [M]$^+$.

HRMS–FAB *(m/z)*: Calc. for [C$_{60}$H$_{58}$N$_4$F]$^+$: 853.4640, found: 853.4641.

IR (ATR, $\tilde{v}$) = 3053 (vw), 2951 (s), 2928 (m), 2901 (m), 2864 (m), 1606 (vs), 1572 (w), 1543 (w), 1513 (m), 1489 (vs), 1473 (vs), 1452 (vs), 1394 (m), 1363 (vs), 1322 (vs), 1295 (vs), 1259 (vs), 1228 (vs), 1201 (m), 1180 (m), 1146 (m), 1092 (vs), 1033 (m), 877 (s), 841 (m), 807 (vs), 793 (s), 752 (s), 738 (vs), 717 (m), 650 (m), 608 (vs), 561 (m), 469 (m), 421 (s) cm^{-1}.

EA: Calc. for C$_{60}$H$_{57}$FN$_4$: C 84.47; H 6.73; N 6.57. Found: C 84.35; H 6.87; N 6.44%.

R$_f$ = 0.66 (Cyclohexane : ethyl acetate = 50:1).

M.p.: 266 °C.

10,10'-(10-(*Tert*-butoxy)dibenzo[*a,c*]phenazine-3,6-diyl)bis(10*H*-phenoxazine) (40a)

A sealable vial was charged with 10,10'-(10-fluorodibenzo[*a,c*]phenazine-3,6-diyl)bis(10*H*-phenoxazine) (365 mg, 552 μmol, 1.00 equiv.), NaO*t*Bu (531 mg, 5.52 mmol, 10.0 equiv.) and Pd(OAc)$_2$ (62.0 mg, 276 μmol, 0.50 equiv.). It was then evacuated and backfilled with argon. Subsequently, toluene (9.86 mL, abs.) was added and P(*t*Bu)$_3$ (112 mg, 552 μL, 552 μmol, 1.00 M in toluene, 1.00 equiv.) was added dropwise while stirring. The reaction mixture was stirred at 110 °C for 6 d. After cooling to room temperature, the mixture was pulled over a Celite® plug with dichloromethane. The filtrate was concentrated under reduced pressure and the crude product was further purified employing column chromatography (cyclohexane : dichloromethane = 5:1 → 2:1). The product (134 mg, 187 μmol, 34% yield) was obtained as an orange-red solid.

^{1}H NMR (400 MHz, CDCl$_3$) δ = 9.72 (d, ^3J = 8.5 Hz, 1H, 8-H), 9.68 (d, ^3J = 8.5 Hz, 1H, 17-H), 8.47–8.53 (m, 2H, 11-H, 14-H), 8.15 (dd, ^3J = 8.6 Hz, ^4J = 1.3 Hz, 1H, 4-H), 7.85 (dd, ^3J = 8.6 Hz, ^3J = 7.6 Hz, 1H, 3-H), 7.73–7.82 (m, 2H, 9-H, 16-H), 7.57 (dd, ^3J = 7.6 Hz, ^4J = 1.3

Hz, 1H, 2-H), 6.74 (ddd, ^3J = 7.9 Hz, ^4J = 1.7 Hz, J = 0.8 Hz, 4H, 2'-H), 6.67 (tdd, ^3J = 7.9 Hz, ^4J = 1.6 Hz, J = 0.9 Hz, 4H, 3'-H), 6.59 (tt, ^3J = 7.9 Hz, ^4J = 1.7 Hz, 4H, 4'-H), 6.05 (ddd, ^3J = 7.9 Hz, J = 3.7 Hz, ^4J = 1.6 Hz, 4H, 5'-H), 1.65 (s, 9H, CH_3) ppm.

^{13}C NMR (101 MHz, CDCl$_3$) δ = 152.50 (C-1), 144.13 (C-10), 144.12 (C-15), 143.76 (4C, C-21), 141.62 (C-5), 141.36 (C-6), 141.24 (C-19), 139.53 (C-20), 134.26 (C-12), 134.24 (C-13), 134.04 (4C, C-26), 131.21 (C-9), 131.10 (C-16), 130.89 (C-18), 130.48 (C-18), 130.27 (C-3), 129.86 (C-8), 129.78 (C-17), 125.91 (C-11), 125.85 (C-14), 124.81 (C-4), 123.46 (4C, C-24), 122.95 (C-2), 121.83 (4C, C-23), 115.80 (4C, C-22), 113.45 (4C, C-25), 81.62 (C$_q$), 29.38 (3C, CH_3) ppm.

MS (FAB, 3-NBA), m/z (%): 715 (23) [M+H]$^+$, 714 (26) [M]$^+$.

IR (ATR, ṽ) = 3055 (w), 1626 (w), 1602 (w), 1592 (w), 1485 (vs), 1463 (s), 1404 (w), 1354 (s), 1333 (vs), 1323 (vs), 1292 (m), 1272 (vs), 1225 (w), 1197 (m), 1128 (w), 1118 (w), 1103 (w), 1091 (m), 1041 (m), 928 (w), 864 (w), 752 (m), 730 (vs), 680 (m), 630 (vs), 611 (w), 592 (w), 548 (w), 482 (w), 452 (m), 441 (m) cm^{-1}.

R$_f$ = 0.76 (Cyclohexane : ethyl acetate = 4:1).

3,6-bis(9,9-dimethylacridin-10(9H)-yl)-10-fluorodibenzo[a,c]phenazine (39b) and **10-(*tert*-butoxy)-3,6-bis(9,9-dimethylacridin-10(9H)-yl)dibenzo[a,c]phenazine (40b)**

A sealable vial was charged with 3,6-dibromo-10-fluorodibenzo[a,c]phenazine (145 mg, 318 µmol, 1.00 equiv), 9,9-dimethyl-10H-acridine (133 mg, 636 µmol, 2.00 equiv), NaOtBu (91.7 mg, 954 µmol, 3.00 equiv), Pd(OAc)$_2$ (7.14 mg, 31.8 µmol, 0.100 equiv) and P(tBu)$_3$ (12.9 mg, 15.4 µL, 63.6 µmol, 1.00 M in toluene, 0.200 equiv). Then, the vial was sealed, evacuated, and backfilled with argon three times. Subsequently, toluene (2.90 mL, abs.) was added and the solution was heated at 110 °C for 21 h. Then, the product mixture was cooled to room temperature and pulled over a Celite® plug with dichloromethane. The filtrate was concentrated under reduced pressure. The crude product was purified *via* column chromatography (cyclohexane : dichloromethane = 4:1 → 2:1). 3,6-bis(9,9-dimethylacridin-10(9H)-yl)-10-fluorodibenzo[a,c]phenazine (203 mg, 285 µmol, 90% yield) was obtained as an orange solid and 10-(*tert*-butoxy)-3,6-bis(9,9-dimethylacridin-10(9H)-yl)dibenzo[a,c]-

phenazine (25.0 mg, 32.6 µmol, 10% yield) was obtained as an orange-yellow solid. The ^{13}C NMR analysis of the latter revealed rotamers, which were indicated using a prime symbol (e.g., C').

^{19}**F NMR** (376 MHz, CDCl$_3$) δ = -124.37 ppm.

^{1}**H NMR** (500 MHz, CD$_2$Cl$_2$) δ = 9.73 (d, $^3J_{H,H}$ = 8.5 Hz, 1H, 8-H), 9.71 (d, $^3J_{H,H}$ = 8.5 Hz, 1H, 17-H), 8.48 (t, $^4J_{H,H}$ = 2.5 Hz, 2H, 11-H, 14-H), 8.23 (dt, $^3J_{H,H}$ = 8.5 Hz, $^4J_{H,H}$ = 1.1 Hz, 1H, 4-H), 7.89 (ddd, $^3J_{H,H}$ = 8.7 Hz, $^3J_{H,H}$ = 7.8 Hz, $^3J_{H,F}$ = 5.4 Hz, 1H, 3-H), 7.78 (ddd, $^3J_{H,H}$ = 8.5 Hz, $^4J_{H,H}$ = 2.5 Hz, 2H, 9-H, 16-H), 7.63 (ddd, $^3J_{C,F}$ = 10.0 Hz, $^3J_{H,H}$ = 7.8 Hz, $^4J_{H,H}$ = 1.1 Hz, 1H, 2-H), 7.50–7.44 (m, 4H, 5'-H), 6.98–6.86 (m, 8H, 3'-H, 4'-H), 6.41–6.31 (m, 4H, 2'-H), 1.69 (s, 12H, CH$_3$) ppm.

^{13}**C NMR** (126 MHz, CD$_2$Cl$_2$) δ = 157.88 (d, $^1J_{C,F}$ = 261.1 Hz, C-1), 144.40 (C-10), 144.35 (C-15), 143.70 (C-6), 143.11 (C-18), 142.27 (d, $^3J_{C,F}$ = 2.29 Hz, C-5), 141.07 (4C, C-1'), 134.91 (C-12), 134.76 (C-13), 133.75 (d, $^2J_{C,F}$ = 12.36 Hz, C-20), 132.20 (C-9), 132.11 (C-16), 130.59 (2C, C-6'), 130.57 (2C, C-6'), 130.34 (C-7), 130.24 (C-18), 130.00 (d, $^3J_{C,F}$ = 8.14 Hz, C-3), 129.95 (C-8), 129.82 (C-17), 126.79 (4C, C-4'), 126.67 (C-11), 126.63 (C-14), 125.85 (4C, 5'), 125.80 (d, $^5J_{C,F}$ = 4.57 Hz, C-4), 121.20 (2C, C-3'), 121.19 (2C, C-3'), 114.51 (4C, C-2'), 114.16 (d, $^2J_{C,F}$ = 18.28 Hz, C-2), 36.35 (2C, C$_q$), 31.72 (4C, CH$_3$) ppm.

MS (FAB, 3-NBA), m/z (%): 714 (42) [M+H]$^+$, 713 (77) [M]$^+$.

HRMS–FAB *(m/z)*: Calc. for [C$_{50}$H$_{38}$N$_4$F]$^+$:713.3075, found 713.3077.

IR (ATR, ṽ) = 3072 (w), 3048 (w), 3040 (w), 2965 (w), 2918 (w), 2857 (w), 1589 (s), 1502 (w), 1480 (s), 1465 (s), 1443 (vs), 1356 (s), 1330 (vs), 1286 (s), 1265 (vs), 1247 (s), 1218 (m), 1205 (w), 1089 (s), 1061 (w), 1047 (m), 931 (w), 849 (w), 741 (vs), 728 (vs), 698 (s), 681 (m), 642 (vs), 568 (m), 523 (m), 487 (m) cm^{-1}.

EA: Calc. for C$_{50}$H$_{37}$FN$_4$: C 84.24; H 5.23; N 7.86. Found: C 84.04; H 5.12; N 7.84%.

R$_f$ = 0.59 (Cyclohexane : ethyl acetate = 4:1).

168

M.p.: 362 °C.

¹H NMR (400 MHz, CDCl₃) δ = 9.77 (d, ^3J = 8.5 Hz, 1H, 8-H), 9.74 (d, ^3J = 8.4 Hz, 1H, 17-H), 8.44 (t, ^4J = 2.1 Hz, 2H, 11-H, 14-H), 8.19 (dd, ^3J = 8.6 Hz, ^4J = 1.3 Hz, 1H, 4-H), 7.87 (dd, ^3J = 8.6 Hz, ^3J = 7.6 Hz, 1H, 3-H), 7.83–7.74 (m, 2H, 9-H, 16-H), 7.58 (dd, ^3J = 7.6 Hz, ^4J = 1.3 Hz, 1H, 2-H), 7.51–7.45 (m, 4H, 5'-H), 7.02–6.87 (m, 8H, 3'-H, 4'-H), 6.41–6.31 (m, 4H, 2'-H), 1.74 (s, 12H, CH_3), 1.67 (s, 9H, OCCH_3) ppm.

¹³C NMR (101 MHz, CDCl₃) δ = 152.54 (C-O), 143.77 (C-10), 143.65 (C-15), 143.50 (C-5), 141.85 (C-6), 141.32 (C-19), 140.85 (2C, C-1'), 140.81 (2C, C-1'), 139.56 (C-20), 134.42 (C-12), 134.22 (C-13), 131.87 (C-9), 131.78 (C-16), 130.74 (C-7), 130.33 (C-18), 130.23 (2C, C-6'), 130.19 (2C, C-6'), 130.16 (4C, C-3), 129.60 (C-8), 129.51 (C-17), 126.61 (4C, C-3'), 126.24 (C-11), 126.16 (C-14), 125.69 (2C, C-5'), 125.63 (2C, C-5'), 124.87 (C-4), 122.95 (C-2), 120.93 (4C, C-4'), 114.22 (2C, C-2'), 114.19 (2C, C-2'), 81.61 (OC(CH₃)₂), 36.17 (2C, C(CH₃)₃), 31.85 (3C$_a$, C(CH₃)₃), 31.75 (3C$_b$, C(CH₃)₃), 29.40 (3C, OC(CH₃)₂) ppm.

MS (FAB, 3-NBA), m/z (%): 767 (62) [M+H]⁺, 751 (51) [M-CH₃]⁺.

HRMS–FAB *(m/z)*: Calc. for [C₅₄H₄₇ON₄]⁺: 767.3744, found: 767.3743.

R$_f$ = 0.82 (Cyclohexane : ethyl acetate = 4:1).

10-(*tert*-butoxy)-3,6-bis(3,6-di-*tert*-butyl-9*H*-carbazol-9-yl)dibenzo[*a,c*]phenazine   (39d) and 3,6,10-tris(3,6-di-*tert*-butyl-9*H*-carbazol-9-yl)dibenzo[*a,c*]phenazine (40c)

A sealable vial was charged with 3,6-dibromo-10-fluorodibenzo[*a,c*]phenazine (200 mg, 438 μmol, 1.00 equiv.), 3,6-di-*tert*-butyl-9*H*-carbazole (245 mg, 877 μmol, 2.00 equiv.), NaO*t*Bu(126 mg, 1.32 mmol, 3.00 equiv.) and Pd(OAc)₂ (9.84 mg, 43.8 μmol, 0.10 equiv.). It was sealed, then evacuated and backfilled with argon three times. Subsequently, toluene (4.00 mL, abs.) was added and P(*t*Bu)₃ (17.7 mg, 21.3 μL, 87.7 μmol, 1.00 M in toluene, 0.20 equiv.) was added dropwise. The resulting reaction mixture was stirrred at 110 °C for 25 h.

After cooling to room temperature, the mixture was pulled over a Celite plug with dichloromethane and concentrated under reduced pressure. The crude mixture was purified by means of column chromatography (cyclohexane : dichloromethane = 5:1 → 4:1). 3,6,10-tris(3,6-di-*tert*-butyl-9*H*-carbazol-9-yl)dibenzo[*a,c*]phenazine (40.0 mg, 36.0 µmol, 8% yield) was obtained as an orange solid, while 10-(*tert*-butoxy)-3,6-bis(3,6-di-*tert*-butyl-9*H*-carbazol-9-yl)dibenzo[*a,c*]phenazine (159 mg, 175 µmol, 40% yield) was obtained as a bright yellow film.

¹H NMR (400 MHz, CDCl₃) δ = 9.66 (d, ^3J = 8.6 Hz, 1H, 8-H), 8.61 (d, ^4J = 2.0 Hz, 1H, 11-H), 8.54 (d, ^4J = 2.0 Hz, 1H, 14-H), 8.51 (dd, ^3J = 8.5 Hz, ^4J = 1.5 Hz, 1H, 4-H), 8.42 (d, ^3J = 8.6 Hz, 1H, 17-H), 8.28 (d, ^4J = 2.0 Hz, 2H, 5'-H), 8.19–8.08 (m, 6H, 2-H, 3-H, 5"-H), 8.02 (dd, ^3J = 8.5 Hz, ^4J = 2.0 Hz, 1H, 9-H), 7.55–7.51 (m, 1H, 16-H), 7.50 (s, 2H, 3"-H), 7.47 (dd, ^3J = 8.7 Hz, ^4J = 1.9 Hz, 2H, 3"-H), 7.43 (dd, ^3J = 8.7 Hz, ^4J = 2.0 Hz, 2H, 3'-H), 7.39–7.38 (m, 4H, 2"-H), 7.23 (d, ^3J = 8.7 Hz, 2H, 2'-H), 1.49 (s, 18H, C*H*₃), 1.45 (s, 18H, C*H*₃), 1.42 (s, 18H, C*H*₃) ppm.

¹³C NMR (101 MHz, CDCl₃) δ = 143.67, 143.57, 143.45, 143.10, 142.54, 141.53, 140.87, 140.65, 138.93, 138.82, 138.43, 136.20, 133.48, 133.06, 129.96, 129.22, 129.00, 128.79, 128.72, 128.65, 128.59, 126.63, 126.43, 124.15, 124.05, 123.93, 123.84, 123.82, 123.60, 120.47, 120.00, 116.57, 116.50, 116.15, 110.62, 109.27, 109.21, 34.95, 34.91, 34.87, 32.23, 32.12, 32.09 ppm.

MS (FAB, 3-NBA), m/z (%): 1130 (71) [M+NH₄]⁺, 1113 (93) [M+H]⁺, 1112 (100) [M]⁺.

IR (ATR, ṽ) = 2952 (s), 2902 (m), 2866 (w), 1608 (s), 1489 (vs), 1472 (vs), 1452 (vs), 1392 (s), 1363 (vs), 1323 (vs), 1295 (vs), 1259 (vs), 1232 (vs), 877 (m), 809 (vs), 758 (s), 612 (s) cm⁻¹.

R$_f$ = 0.85 (Cyclohexane : ethyl acetate = 4:1).

¹H NMR (400 MHz, CDCl₃) δ = 9.74 (d, ^3J = 7.9 Hz, 1H, 8-H), 9.70 (d, ^3J = 8.8 Hz, 1H, 17-H), 8.74 (d, ^4J = 2.3 Hz, 2H, 11-H, 14-H), 8.23 (d, ^4J = 1.9 Hz, 4H, 25-H), 8.16 (dd, ^3J = 8.3 Hz, ^4J = 1.3 Hz, 1H, 4-H), 8.09 (ddd, ^3J = 8.8 Hz, ^3J = 7.9 Hz, ^4J = 1.8 Hz, 2H, 9-H, 16-H), 7.85 (t, ^3J = 8.3 Hz, 1H, 3-H), 7.63–7.56 (m, 5H, 2-H, 23-H), 7.56–7.48 (m, 4H, 22-H), 1.72 (s, 9H, OCH₃), 1.52 (s, 36H, CH₃) ppm.

¹³C NMR (101 MHz, CDCl₃) δ = 152.48 (C-1), 143.65 (C-5), 143.59 (2C, C-24), 143.57 (2C, C-24), 141.76 (C-10), 141.16 (C-15), 140.54 (C-6), 140.39 (C-19), 139.42 (C-20), 139.02 (4C, C-21), 133.23 (C-12), 133.04 (C-13), 129.84 (C-3), 129.45 (C-7), 129.04 (C-18), 128.56 (C-8), 128.50 (C-17), 126.69 (C-9), 126.59 (C-16), 124.80 (C-4), 124.14 (4C, C-23), 123.92 (2C, C-26), 123.91 (2C, C-26), 122.75 (C-2), 120.45 (C-11), 120.37 (C-14), 116.55 (2C, C-25), 116.52 (2C, C-25), 109.35 (2C, C-22), 109.31 (2C, C-22), 81.48 (C-28), 34.90 (4C, C-27), 32.13 (12C, CH₃^{tBu}), 29.44 (3C, CH₃^{OtBu}) ppm.

MS (FAB, 3-NBA), m/z (%): 908 (75), 907 (100).

R$_f$ = 0.84 (Cyclohexane : ethyl acetate = 4:1).

3',6'-Dibromodispiro[[1,3]dioxolane-2,9'-phenanthrene-10',2"-[1,3]dioxolane] (41)

A sealable vial was charged with 3,6-dibromophenanthrene-9,10-dione (3.87 g, 10.6 mmol, 1.00 equiv.) and *p*-toluenesulfonic acid monohydrate (4.02 g, 21.1 mmol, 2.00 equiv.). It was then sealed and evacuated, and backfilled with argon three times. Subsequently, toluene (77.4 mL, abs.) and ethylene glycol (13.1 g, 11.8 mL, 211 mmol, 20.00 equiv.) were added dropwise to the solution while stirring. The reaction mixture was refluxed at 140 °C for 6 h. After cooling to room temperature, the mixture was filtered off, and the residue was washed with water, methanol, and diethyl ether. The product (4.07 g, 8.55 mmol, 85%) was obtained as a colorless solid.

¹H NMR (400 MHz, CDCl₃) δ = 7.96 (d, ⁴*J* = 1.9 Hz, 2H, 5-H), 7.62 (d, ³*J* = 8.3 Hz, 2H, 8-H), 7.56 (dd, ³*J* = 8.3 Hz, ⁴*J* = 1.9 Hz, 2H, 7-H), 4.19 (s, 4H, C*H*₂), 3.63 (s, 4H, C*H*₂) ppm.

¹³C NMR (101 MHz, CDCl₃) δ = 134.02 (2C, C-4b), 132.39 (2C, C-7), 132.33 (2C, C-8a), 128.40 (2C, C-8), 127.22 (2C, C-5), 124.58 (2C, C-6), 92.32 (2C, C-9), 61.50 (4C, *C*H₂) ppm.

MS (FAB, 3-NBA), m/z (%): 456 (27) [M(⁸¹Br₂)+H]⁻, 455 (11) [M(⁸¹Br₂)]⁺, 454 (66) [M(⁸¹Br⁷⁹Br)+H]⁺, 453 (6) [M(⁸¹Br⁷⁹Br)]⁺, 452 (34) [M(⁷⁹Br₂)+H]⁻.

HRMS–FAB *(m/z)*: Calc. for [C₁₈H₁₄O₄⁷⁹Br₂]⁺: 451.9253; found: 451.9253.

IR (ATR, ṽ) = 3080 (w), 2979 (w), 2962 (w), 2925 (w), 2873 (w), 1592 (w), 1558 (w), 1477 (w), 1462 (w), 1395 (w), 1387 (w), 1281 (m), 1272 (w), 1242 (w), 1186 (w), 1089 (vs), 1047 (s), 1018 (m), 996 (w), 979 (s), 965 (m), 946 (vs), 931 (m), 888 (s), 870 (s), 824 (m), 813 (vs), 755 (w), 703 (w), 591 (w), 547 (m), 528 (m), 475 (w), 456 (w), 426 (w), 381 (w) cm⁻¹.

EA: Calc. for C₁₈H₁₄O₄⁷⁹Br₂: C 47.61; H 3.11. Found: C 47.45; H 2.78%.

R*f* = 0.47 (Cyclohexane : ethyl acetate = 2:1).

M.p. : 310 °C.

3',6'-Di(10*H*-phenoxazin-10-yl)dispiro[[1,3]dioxolane-2,9'-Phenanthrene-10',2''-[1,3]dioxolane] (42a)

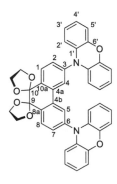

A sealable vial was charged with 3,6-dibromophenanthrene-9,10-dione (1.00 g, 2.20 mmol, 1.00 equiv.), 10*H*-phenoxazine (0.81 g, 4.40 mmol, 2.00 equiv.), [(*t*Bu)₃PH]BF₄ (0.13 g, 0.44 mmol, 0.20 equiv.), NaO*t*Bu (0.64 g, 6.61 mmol, 3.00 equiv.) and Pd(dba)₂ (0.13 g, 0.22 mmol, 0.10 equiv.). It was then evacuated and backfilled with argon. Then, toluene (20.0 mL, abs.) was added. The reaction mixture was stirred at 120 °for 24 h. The product mixture was concentrated under reduced pressure. The crude mixture was further purified employing column chromatography (cyclohexane :

ethyl acetate = 50:1). The product (1.05 g, 1.59 mmol, 72%) was obtained as a pale-yellow solid.

^{1}H NMR (400 MHz, CDCl$_3$) δ = 8.04 (d, ^3J = 8.1 Hz, 2H, 1-H, 8-H), 7.83 (d, ^4J = 2.0 Hz, 2H, 4-H, 5-H), 7.45 (dd, ^3J = 8.1 Hz, ^4J = 2.0 Hz, 2H, 2-H, 7-H), 6.68 (dd, ^3J = 7.6 Hz, ^4J = 1.5 Hz, 4H, 5'-H), 6.63 (td, ^3J = 7.6 Hz, ^4J = 1.5 Hz, 4H, 4'-H), 6.56 (td, ^3J = 7.6 Hz, ^4J = 1.8 Hz, 4H, 2'-H), 5.97 (dd, ^3J = 7.8 Hz, ^3J = 1.6 Hz, 4H, 2'-H), 4.33 (s, 4H, CH_2), 3.81 (s, 4H, CH_2) ppm.

^{13}C NMR (101 MHz, CDCl$_3$) δ = 144.02 (4C, C-6'), 141.13 (2C, C-3, C-6), 135.46 (2C, C-4a, C-4b), 134.05 (4C, C-2'), 133.30 (2C, C-8a, C-10a), 132.02 (2C, C-2, C-7), 129.67 (2C, C-1, C-8), 126.88 (2C, C-4, C-5), 123.40 (4C, C-3'), 121.74 (4C, C-4'), 115.70 (4C, C-5'), 113.50 (4C, C-2'), 92.54 (2C, C-9, C-10), 60.54 (4C, CH_2) ppm.

MS (FAB, 3-NBA), m/z (%): 659 (20) [M+H]$^+$, 658 (33) [M]$^+$.

HRMS–FAB (*m/z*): Calc. for [C$_{42}$H$_{30}$O$_6$N$_2$]$^+$: 658.2098; found: 658.2100.

IR (ATR, ṽ) = 3058 (w), 2975 (w), 2918 (w), 2868 (w), 1740 (w), 1592 (w), 1485 (vs), 1463 (s), 1452 (m), 1330 (vs), 1292 (m), 1269 (vs), 1238 (m), 1200 (m), 1186 (m), 1094 (vs), 1043 (m), 1031 (m), 989 (m), 977 (m), 955 (vs), 936 (m), 924 (m), 894 (w), 863 (m), 745 (vs), 731 (vs), 633 (s), 596 (w), 387 (w) cm^{-1}.

EA: Calc. for C$_{42}$H$_{30}$O$_6$N$_2$: C 76.58; H 4.59; N 4.25. Found C 76.23; H 4.79; N 4.02%.

R$_f$ = 0.40 (Cyclohexane : ethyl acetate = 5:1).

3',6'-Bis(9,9-dimethylacridin-10(9H)-yl)dispiro[[1,3]dioxolane-2,9'-Phenanthrene-10',2''-[1,3]dioxolane] (42b)

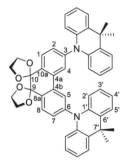

A sealable vial was charged with 3,6-dibromophenanthrene-9,10-dione (1.00 g, 2.20 mmol, 1.00 equiv.), 9,9-dimethyl-10H-acridine (0.92 g, 4.40 mmol, 2.00 equiv.), P(tBu)₃ (0.89 mg, 0.44 mL, 0.44 mmol, 0.20 equiv.), NaOtBu (0.64 g, 6.61 mmol, 3.00 equiv.) and Pd(OAc)₂ (0.50 g, 0.22 mmol, 0.10 equiv.). It was then evacuated and backfilled with argon. Subsequently, toluene (20.0 mL, abs.) was added and the solution was stirred at 120 °C for 24 h. After cooling to room temperature, all volatile components were removed under reduced pressure. The crude mixture was further purified employing column chromatography (cyclohexane : ethyl acetate = 50:1). The product (1.52 g, 2.14 mmol, 97%) was obtained as an orange solid.

¹H NMR (400 MHz, CDCl₃) δ = 8.09 (d, ^3J = 7.9 Hz, 2H, 1-H, 8-H), 7.77 (d, ^4J = 1.8 Hz, 2H, 4-H, 5-H), 7.46 – 7.39 (m, 6H, 2-H, 7-H, 1'-H, 8'-H), 7.00 – 6.81 (m, 8H, 3'-H, 4'-H), 6.29 (dd, ^3J = 7.9 Hz, ^4J = 1.8 Hz, 4H, 2'-H), 4.38 (s, 4H, CH₂), 3.89 (s, 4H, CH₂), 1.67 (s, 12H, CH₃) ppm.

¹³C NMR (101 MHz, CDCl₃) δ = 143.34 (2C, C-3, C-6), 140.54 (2C, C-1'), 135.48 (2C, C-4a, C-4b), 133.00 (2C, C-8a, C-10a), 132.42 (2C, C-3, C-6), 130.01 (4C, C-6'), 129.45 (2C, C-2, C-7), 127.31 (2C, C-4, C-5), 126.56 (4C, C-3'), 125.63 (4C, C-5'), 120.82 (4C, C-4'), 114.20 (4C, C-2'), 92.73 (2C, C-7'), 36.04 (4C, CH₂), 31.88 (4C, CH₃) ppm.

MS (FAB, 3-NBA), m/z (%): 712 (29) [M+H]⁺, 711 (72) [M]⁺.

HRMS–FAB (m/z): Calc. for [C₄₈H₄₂O₄N₂]⁺: 711.3217; found: 711.3216.

IR (ATR, ṽ) = 3030 (vw), 2970 (w), 2956 (w), 2876 (vw), 1728 (vw), 1589 (w), 1502 (w), 1472 (s), 1446 (s), 1324 (s), 1282 (m), 1268 (m), 1215 (w), 1193 (w), 1115 (m), 1094 (vs), 1045 (m), 1031 (w), 990 (w), 980 (m), 956 (s), 929 (w), 894 (w), 844 (w), 742 (vs), 718 (w), 643 (m), 564 (w), 526 (w), 514 (w), 432 (w) cm⁻¹.

EA: Calc. for C₄₈H₄₂O₄N₂: C 81.1; H 5.96; N 3.94. Found: C 80.30; H 5.88; N 3.82%.

R*f* = 0.29 (cyclohexane : ethyl acetate = 5:1).

M.p.: 340 °C.

3',6'-Bis(3,6-di-*tert*-butyl-9*H*-carbazol-9-yl)dispiro[[1,3]dioxolane-2,9'-Phenanthrene-10',2''-[1,3]dioxolane] (42c)

To a sealable vial, 3,6-dibromophenanthrene-9,10-dione (1.00 g, 2.20 mmol, 1.00 equiv.), 3,6-di-*tert*-butyl-9*H*-carbazole (1.23 g, 4.40 mmol, 2.00 equiv.), NaO*t*Bu (0.64 g, 6.61 mmol, 3.00 equiv.), Pd(OAc)$_2$ (0.05 g, 0.22 mmol, 0.10 equiv.) were added, and the vial was evacuated and backfilled with argon. Then, toluene (abs., 20.0 mL) was added and P(*t*Bu)$_3$ (0.09 g, 0.44 mL, 0.44 mmol, 0.20 equiv.) was added dropwise. Subsequently, the solution was stirred at 110 °C for 24 hours. Then, the product mixture was concentrated under reduced pressure. The crude product was purified *via* column chromatography(cyclohexane : ethyl acetate = 50:1). The product (1.70 g, 1.99 mmol, 91%) was obtained as a bright orange-yellow solid.

¹H NMR (400 MHz, CDCl$_3$) δ = 8.13 (m, 4H, 4'-H, 5'-H), 8.08 (d, ^4J = 2.0 Hz, 2H, 4-H, 5-H), 8.05 (d, ^3J = 8.2 Hz, 2H, 1-H, 8-H), 7.70 (dd, ^3J = 8.2 Hz, ^4J = 2.0 Hz, 2H, 2-H, 7-H), 7.43 (m, 8H, 1'-H, 2'-H, 7'-H, 8'-H), 4.37 (s, 4H, C*H*$_2$), 3.88 (s, 4H, C*H*$_2$), 1.46 (s, 36H, C*H*$_3$) ppm.

¹³C NMR (101 MHz, CDCl$_3$) δ = 143.32 (4C, C-4'), 140.25 (2C, C-3, C-6), 138.94 (4C, C-1'), 134.26 (2C, C-4a, C-4b), 131.63 (2C, C-8a, C-10a), 128.31 (2C, C-1, C-8), 127.18 (2C, C-2, C-7), 123.97 (4C, C-3'), 123.67 (4C, C-6'), 121.99 (2C, C-4, C-5), 116.41 (4C, C-5'), 109.31 (4C, C-2'), 92.74 (2C, C-9, C-10), 34.86 (4C, C$_q$), 32.11 (12C, C*H*$_3$) ppm.

MS (FAB, 3-NBA), m/z (%): 851 (80) [M+H]$^+$, 850 (100) [M]$^+$.

HRMS–FAB (*m/z*): Calc. for [C$_{58}$H$_{62}$O$_4$N$_2$]$^+$: 850.4704; found: 850.4705.

IR (ATR, ṽ) = 2952 (m), 2902 (w), 2867 (w), 1609 (w), 1500 (m), 1487 (m), 1472 (s), 1452 (m), 1392 (w), 1363 (m), 1323 (w), 1293 (m), 1276 (m), 1262 (s), 1232 (m), 1203 (w), 1181 (m), 1094 (vs), 1054 (w), 1030 (m), 1001 (w), 979 (s), 955 (vs), 924 (w), 891 (w), 877 (m), 841 (w), 807 (vs), 764 (w), 741 (m), 656 (w), 611 (s), 594 (w), 560 (w), 531 (w), 476 (w), 472 (w), 421 (w), 404 (w), 391 (w), 384 (w), 377 (w) cm^{-1}.

EA: Calc. for $C_{58}H_{62}O_4N_2$: C 81.85; H 7.34; N 3.29. Found: C 81.61; H 7.37; N 3.28%.

R$_f$ = 0.27 (Cyclohexane : ethyl acetate = 5:1).

10,10'-(11,12-difluorodibenzo[a,c]phenazine-3,6-diyl)bis(10H-phenoxazine) (43c)

A sealable vial was charged with 3,6-di(10H-phenoxazin-10-yl)phenanthrene-9,10-dione (50.0 mg, 87.6 μmol, 1.00 equiv.) and 4,5-difluorobenzene-1,2-diamine (12.6 mg, 87.6 μmol, 1.00 equiv.). Then, n-butanol (2.00 mL) was added, and the reaction mixture was stirred at 120 °C for 38 h. The product (47.0 mg, 69.3 μmol, 79% yield) was recrystallized from a solution of the crude mixture in dichloromethane as a red solid.

^{1}H NMR (400 MHz, CDCl$_3$) δ = 9.62 (d, $^3J_{H,H}$ = 8.5 Hz, 2H, 6-H), 8.49 (d, $^4J_{H,H}$ = 2.0 Hz, 2H, 9-H), 8.12 (t, $^3J_{H,F}$ = 9.3 Hz, 2H, 2-H), 7.78 (dd, $^3J_{H,H}$ = 8.5 Hz, $^4J_{H,H}$ = 1.9 Hz, 2H, 7-H), 6.73 (dd, $^3J_{H,H}$ = 7.8 Hz, $^4J_{H,H}$ = 1.6 Hz, 4H, 2'-H), 6.67 (td, $^3J_{H,H}$ = 7.8 Hz, $^4J_{H,H}$ = 1.5 Hz, 4H, 3'-H), 6.58 (td, $^3J_{H,H}$ = 7.8 Hz, $^4J_{H,H}$ = 1.6 Hz, 4H, 4'-H), 6.02 (dd, $^3J_{H,H}$ = 7.8 Hz, $^4J_{H,H}$ = 1.6 Hz, 4H, 5'-H) ppm.

MS (ESI), m/z (%): 679 (48) [M+H]$^+$, 678 (100) [M]$^+$.

IR (ATR, ṽ) = 3046 (w), 1605 (w), 1589 (w), 1482 (vs), 1463 (s), 1405 (w), 1354 (m), 1333 (s), 1315 (s), 1292 (s), 1272 (vs), 1230 (m), 1196 (s), 1153 (m), 1118 (m), 1038 (m), 878 (m), 864 (s), 837 (m), 739 (vs), 730 (vs), 708 (s), 677 (m), 629 (vs), 611 (m), 486 (w), 441 (m) cm^{-1}.

R$_f$ = 0.79 (Cyclohexane : ethyl acetate = 4:1).

10,10'-(11,12-dimethyldibenzo[*a,c*]phenazine-3,6-diyl)bis(10*H*-phenoxazine) (43d)

A sealable vial was charged with 3,6-di(10*H*-phenoxazin-10-yl)phenanthrene-9,10-dione (50.0 mg, 87.6 μmol, 1.00 equiv.) and 4,5-dimethylbenzene-1,2-diamine (11.9 mg, 87.6 μmol, 1.00 equiv.). It was sealed, and *n*-butanol (2.00 mL) was added. The reaction mixture was stirred at 120 °C for 38 h. After cooling to room temperature, the crude mixture was pulled over a silica plug with dichloromethane. Subsequently, all volatile components were removed, and the product (38.0 mg, 56.7 μmol, 65% yield) was recrystallized from a cyclohexane/ethyl acetate mixture as a yellow-orange solid. Due to difficulties in solubility of the substrate, the characterization by means of ^{13}C NMR was not effective as the high signal-to-noise ratio impeded the differentiation of signals indicated by HSQC and HMBC from the background.

^{1}H NMR (400 MHz, CDCl$_3$) δ = 9.66 (d, ^3J = 8.5 Hz, 2H, 6-H), 8.47 (d, ^4J = 1.9 Hz, 2H, 9-H), 8.15 (s, 2H, 2-H), 7.76 (dd, ^3J = 8.5 Hz, ^4J = 1.9 Hz, 2H, 7-H), 6.72 (dd, ^3J = 7.8 Hz, ^4J = 1.6 Hz, 4H, 5'-H), 6.66 (td, ^3J = 7.8 Hz, ^4J = 1.5 Hz, 4H, 4'-H), 6.57 (td, ^3J = 7.8 Hz, ^4J = 1.6 Hz, 4H, 3'-H), 6.03 (dd, ^3J = 7.8 Hz, ^4J = 1.5 Hz, 4H, 2'-H), 2.73–2.47 (m, 6H, C*H$_3$*) ppm.

MS (ESI), m/z (%): 671 (55) [M+H]$^+$, 670 (100) [M]$^+$.

IR (ATR, ṽ) = 3058 (w), 1604 (w), 1589 (w), 1483 (vs), 1463 (m), 1405 (w), 1357 (m), 1329 (s), 1313 (s), 1290 (m), 1272 (vs), 1193 (m), 1164 (w), 1116 (w), 1058 (w), 1041 (w), 936 (w), 926 (w), 921 (w), 871 (w), 863 (m), 839 (m), 752 (m), 734 (vs), 710 (m), 680 (w), 630 (s), 615 (w), 606 (w), 484 (w), 459 (w), 443 (w), 397 (w) cm^{-1}.

EA: Calc. for C$_{46}$H$_{30}$N$_4$O$_2$: C 82.37; H 4.51; N 8.35. Found: C 80.25; H 4.43; N 8.01%.

R$_f$ = 0.62 (Cyclohexane : ethyl acetate = 4:1).

(3,6-di(10*H*-phenoxazin-10-yl)dibenzo[*a,c*]phenazin-11-yl)(phenyl)methanone (43e)

A sealable vial was charged with 3,6-di(10H-phenoxazin-10-yl)phenanthrene-9,10-dione (50.0 mg, 87.6 µmol, 1.00 equiv.) and (3,4-diamino-phenyl)-phenylmethanone (18.6 mg, 87.6 µmol, 1.00 equiv.). Then, it was sealed and *n*-butanol (2.00 mL) was added. The reaction mixture was stirred at 120 °C for 16 h. Subsequently, all volatile components were removed under reduced pressure and the violet crude product was purified employing column chromatography (cyclohexane : dichloromethane = 4:1). The product (62.0 mg, 83.0 µmol, 95% yield) was obtained as a dark violet-red solid.

¹H NMR (400 MHz, CDCl$_3$) δ = 9.72 (d, ^3J = 8.5 Hz, 1H, 7-H), 9.64 (d, ^3J = 8.5 Hz, 1H, 16-H), 8.75 (d, ^4J = 1.9 Hz, 1H, 20-H), 8.52–8.49 (m, 3H, 3-H, 10-H, 13-H), 8.41 (dd, ^3J = 8.8 Hz, ^4J = 1.9 Hz, 1H, 2-H), 8.04–7.96 (m, 2H, 2'-H), 7.81 (dd, ^3J = 8.5 Hz, $4J$ = 1.9 Hz, 1H, 8-H), 7.77 (dd, ^3J = 8.5 Hz, ^4J = 1.9 Hz, 1H, 15-H), 7.74–7.67 (m, 1H, 4'-H), 7.60 (dd, ^3J = 8.4 Hz, ^3J = 7.0 Hz, 2H, 3'-H), 6.73 (ddd, ^3J = 7.8 Hz, J = 3.4 Hz, ^4J = 1.6 Hz, 4H, 5"-H), 6.67 (tdd, ^3J = 7.8 Hz, J = 3.3 Hz, ^4J = 1.5 Hz, 4H, 4"-H), 6.58 (tdd, ^3J = 7.8 Hz, J = 3.9 Hz, ^4J = 1.6 Hz, 4H, 3"-H), 6.04 (td, ^3J = 7.8 Hz, ^4J = 1.5 Hz, 4H, 2"-H) ppm.

¹³C NMR (101 MHz, CDCl$_3$) δ = 195.93 (C=O), 144.14 (5C, C-4, C-6"), 143.33 (C-5), 142.97 (C-18), 142.10 (C-9), 141.89 (C-19), 141.58 (C-14), 138.78 (C-1), 137.33 (C-1'), 134.58 (C-11), 134.29 (C-12), 134.14 (4C, C-1"), 133.18 (C-4'), 133.00 (C-20), 131.40 (2C, C-8, C-15), 130.40 (2C, C-2'), 130.27, 130.24, 130.21, 130.16, 130.15, 129.90 (C-16), 128.80 (2C, C-3'), 126.06 (2C, C-10, C-13), 123.47 (4C, C-3"), 121.98 (2C, C-5"), 121.95 (2C, C-5"), 115.90 (2C, C-4"), 115.88 (2C, C-4"), 113.44 (4C, C-2") ppm.

MS (FAB, 3-NBA), m/z (%): 747 (49) [M+H]$^+$, 746 (37) [M]$^+$.

HRMS–FAB *(m/z)*: Calc. for [C$_{51}$H$_{30}$O$_3$N$_4$]$^+$: 746.2312; found: 746.2309.

IR (ATR, ṽ) = 1655 (w), 1596 (m), 1485 (vs), 1465 (s), 1401 (w), 1360 (s), 1334 (vs), 1292 (m), 1266 (vs), 1221 (m), 1204 (m), 1198 (m), 1180 (m), 1119 (m), 1054 (m), 1044 (m), 884 (m), 728 (vs), 707 (s), 700 (vs), 629 (vs), 603 (m), 442 (m) cm^{-1}.

R_f = 0.79 (Cyclohexane : ethyl acetate = 2:1).

3,6-bis(9,9-dimethylacridin-10-yl)-11,12-difluoro-phenanthro[9,10-*b*]quinoxaline (44c)

A sealable vial was charged with 3,6-bis(9,9-dimethylacridin-10(9*H*)-yl)phenanthrene-9,10-dione (50.0 mg, 80.3 µmol, 1.00 equiv.) and 4,5-difluorobenzene-1,2-diamine (12.6 mg, 87.6 µmol, 1.09 equiv.). Then, the vial was sealed and *n*-butanol (2.00 mL) was added. The reaction mixture was stirred at 120 °C for 45 h. Then, all volatile components were removed and the crude mixture was further purified through column chromatography using a cyclohexane/ethyl acetate gradient. The product (48.0 mg, 65.7 µmol, 82% yield) was obtained as a yellow solid. Characterization by means of ^{13}C NMR was not effective due to the low solubility of the substrate. Because of the resulting high signal-to-noise ratio, some signals indicated by means of HSQC and HMBC NMR could not be distinguished from the background signals.

^{1}H NMR (400 MHz, CDCl$_3$) δ = 9.67 (d, $^3J_{H,H}$ = 8.5 Hz, 2H, 6-H), 8.43 (d, $^4J_{H,H}$ = 1.9 Hz, 2H, 9-H), 8.14 (t, $^3J_{H,F}$ = 9.3 Hz, 2H, 2-H), 7.77 (dd, $^3J_{H,H}$ = 8.5 Hz, $^4J_{H,H}$ = 1.9 Hz, 2H, 7-H), 7.50–7.43 (m, 4H, 5'-H), 6.97–6.86 (m, 8H, 3'-H, 4'-H), 6.35–6.27 (m, 4H, 2'-H), 1.72 (s, 12H, CH_3) ppm.

R_f = 0.83 (Cyclohexane : ethyl acetate = 4:1).

3,6-bis(9,9-dimethylacridin-10(9H)-yl)-11,12-dimethyldibenzo[a,c]phenazine (44d)

A sealable vial was charged with 3,6-di(10H-phenoxazin-10-yl)phenanthrene-9,10-dione (50.0 mg, 87.6 µmol, 1.00 equiv.) and 4,5-dimethylbenzene-1,2-diamine (11.9 mg, 87.6 µmol, 1.00 equiv.). It was sealed, and n-butanol (2.00 mL) was added. The reaction mixture was stirred at 120 °C for 3 h. The precipitate was filtered off, washed with methanol, and dried under reduced pressure to obtain the product (52.0 mg, 71.9 µmol, 90% yield) as a bright yellow solid.

^{1}H NMR (400 MHz, CDCl$_3$) δ = 9.71 (d, ^3J = 8.5 Hz, 2H, 6-H), 8.43 (d, ^4J = 1.9 Hz, 2H, 9-H), 8.17 (s, 2H, 2-H), 7.75 (dd, ^3J = 8.5 Hz, ^4J = 1.9 Hz, 2H, 7-H), 7.51–7.44 (m, 4H, 5'-H), 6.97–6.88 (m, 8H, 3'-H, 4'-H), 6.37–6.31 (m, 4H, 2'-H), 2.65 (s, 6H, CH$_3$), 1.73 (s, 12H, CH$_3$) ppm.

^{13}C NMR (101 MHz, CDCl$_3$) δ = 143.21 (2C, C-8), 141.84 (2C, C-3), 141.59 (2C, C-1), 141.39 (2C, C-4), 140.86 (4C, C-1'), 133.99 (2C, C-10), 131.67 (2C, C-9), 130.66 (2C, C-5), 130.16 (4C, C-6'), 129.22 (2C, C-6), 128.51 (2C, C-2), 126.60 (4C, C-3'), 126.17 (2C, C-9), 125.66 (4C, C-2'), 120.88 (4C, C-4'), 114.20 (4C, C-5'), 36.16 (2C, C$_q$), 31.85 (4C, CH$_3$), 20.82 (2C, CH$_3$) ppm.

MS (FAB, 3-NBA), m/z (%): 723 (5) [M+H]$^+$.

HRMS–FAB (m/z): Calc. for [C$_{52}$H$_{43}$N$_4$]$^+$: 723.3482; found: 723.3479.

IR (ATR, ṽ) = 3070 (w), 3034 (w), 2952 (w), 2917 (w), 2854 (w), 1591 (m), 1470 (s), 1446 (vs), 1358 (m), 1333 (s), 1323 (vs), 1268 (vs), 1057 (w), 1045 (m), 739 (vs), 714 (m), 640 (s), 568 (m) cm^{-1}.

R$_f$ = 0.86 (Cyclohexane : ethyl acetate = 2:1).

(3,6-bis(9,9-dimethylacridin-10(9*H*)-yl)dibenzo[*a*,*c*]phenazin-11-yl)(phenyl)methanone (44e)

A sealable vial was charged with 3,6-bis(9,9-dimethylacridin-10(9H)-yl)phenanthrene-9,10-dione (50.0 mg, 80.3 µmol, 1.00 equiv.) and (3,4-diaminophenyl)-phenylmethanone (18.6 mg, 87.6 µmol, 1.00 equiv.). Then, it was sealed, and n-butanol (2.00 mL) was added. The reaction mixture was stirred at 120 °C for 2 h. Subsequently, the precipitate is filtered off, washed with methanol, and dried under reduced pressure to give the product (54.0 mg, 67.6 µmol, 84% yield) as a bright orange-red solid.

^{1}H NMR (400 MHz, CDCl$_3$) δ = 9.77 (d, ^3J = 8.5 Hz, 1H, 15-H), 9.69 (d, ^3J = 8.5 Hz, 1H, 6-H), 8.78 (d, ^4J = 1.8 Hz, 1H, 2-H), 8.53 (d, ^3J = 8.8 Hz, 1H, 19-H), 8.45–8.41 (m, 3H, 9-H, 12-H, 20-H), 8.06–7.98 (m, 2H, 2'-H), 7.80 (dd, ^3J = 8.5 Hz, ^4J = 1.8 Hz, 1H, 14-H), 7.76 (dd, ^3J = 8.5 Hz, ^4J = 1.8 Hz, 1H, 7-H), 7.74–7.68 (m, 1H, 4'-H), 7.61 (dd, ^3J = 8.3 Hz, ^3J = 7.0 Hz, 2H, 3'-H), 7.51–7.44 (m, 4H, 5"-H), 6.97–6.89 (m, 8H, 3"-H, 4"-H), 6.37–6.29 (m, 4H, 2"-H), 1.73 (s, 6H, C*H*$_3$), 1.72 (s, 6H, C*H*$_3$) ppm.

^{13}C NMR (101 MHz, CDCl$_3$) δ = 196.00 (C=O), 144.40 (C-13), 144.17 (C-8), 144.14 (C-18), 143.58 (C-17), 143.20 (C-4), 141.56 (C-3), 140.74 (4C, C-1"), 138.68 (C-1), 137.38 (C-1'), 134.78 (C-11), 134.48 (C-10), 133.15 (C-4'), 133.07 (C-2), 132.06 (2C, C-7, C-14), 130.42 (2C, C-2'), 130.29 (C-20), 130.26 (4C, C-6"), 130.18 (C-19), 130.00 (2C, C-5, C-16), 129.95 (C-15), 129.64 (C-6), 128.80 (2C, C-3'), 126.62 (4C, C-3"), 126.36 (2C, C-9, C-12), 125.74 (4C, C-2"), 121.05 (2C, C-4"), 121.02 (2C, C-4"), 114.19 (2C, C-5"), 114.16 (2C, C-5"), 36.17 (2C, C$_q$), 31.83 (4C, C*H*$_3$) ppm.

MS (FAB, 3-NBA), m/z (%): 799 (100) [M+H]$^+$, 783 (80) [M-CH$_3$]$^+$.

HRMS–FAB *(m/z)*: Calc. for [C$_{57}$H$_{43}$ON$_4$]$^+$: 799.3431; found: 799.3434.

IR (ATR, ṽ) = 1656 (w), 1589 (m), 1472 (m), 1445 (s), 1357 (m), 1326 (s), 1259 (vs), 1048 (m), 890 (w), 742 (vs), 710 (s), 640 (s), 568 (w) cm^{-1}.

R_f = 0.76 (Cyclohexane : ethyl acetate = 4:1).

3,6-bis(9,9-dimethylacridin-10(9H)-yl)dibenzo[a,c]phenazine-11-sulfonic acid (44f)

A sealable vial was charged with 3,6-di(10H-phenoxazin-10-yl)phenanthrene-9,10-dione (50.0 mg, 87.6 µmol, 1.00 equiv.) and 3,4-diaminobenzene-sulfonic acid (16.5 mg, 87.6 µmol, 1.00 equiv.). Then, n-butanol (2.00 mL) was added and the reaction was stirred at 120 °C for 49 h. The product mixture was concentrated and recrystallized from a cyclohexane and ethyl acetate (50:1) mixture. Subsequently, the precipitate was washed with methanol three times to give the product (42.0 mg, 54.2 µmol, 68% yield) as a rust-red solid.

^{1}H NMR (400 MHz, DMSO-d$_6$) δ = 9.66 (dd, ^3J = 9.6 Hz, ^3J = 8.5 Hz, 2H, 6-H, 15-H), 8.99 (t, ^4J = 2.4 Hz, 2H, 9-H, 12-H), 8.57 (d, ^4J = 1.9 Hz, 1H, 2-H), 8.41 (d, ^3J = 8.8 Hz, 1H, 19-H), 8.23 (dd, ^3J = 8.8 Hz, ^4J = 1.9 Hz, 1H, 20-H), 7.81 (ddd, ^3J = 8.6 Hz, J = 5.3 Hz, ^4J = 1.8 Hz, 2H, 7-H, 14-H), 7.50 (dd, ^3J = 7.5 Hz, ^4J = 1.7 Hz, 4H, 5'-H), 6.93 (ddt, ^3J = 8.4 Hz, ^3J = 7.5 Hz, ^4J = 1.7 Hz, 4H, 3'-H), 6.87 (td, ^3J = 7.5 Hz, ^4J = 1.3 Hz, 4H, 4'-H), 6.20 (ddd, ^3J = 8.4 Hz, J = 3.5 Hz, ^4J = 1.3 Hz, 4H, 2'-H), 1.66 (s, 12H, CH$_3$) ppm.

IR (ATR, ṽ) = 3403 (m), 3075 (w), 3031 (w), 2970 (w), 1591 (s), 1504 (w), 1473 (s), 1443 (vs), 1358 (s), 1329 (vs), 1266 (s), 1187 (s), 1094 (vs), 1031 (vs), 931 (m), 832 (m), 741 (vs), 714 (vs), 683 (vs), 664 (s), 652 (vs), 640 (vs), 622 (vs), 613 (vs), 602 (vs), 581 (vs), 568 (vs), 547 (vs), 518 (vs), 490 (vs), 460 (vs), 452 (vs), 439 (vs), 426 (vs), 408 (vs), 395 (vs), 382 (vs) cm^{-1}.

MS (ESI, neg.), m/z (%): 774 (57) [M]$^-$, 773 (100) [M-H]$^-$.

R_f = 0.06 (Dichloromethane : methanol = 10:1).

182

5.2.2. Pyrrolo[3,4-*f*]isoindole-1,3,5,7(2*H*,6*H*)-tetraone (PIT)-Based Molecule Design

4,8-Dibromo-2,6-dimesitylpyrrolo[3,4-*f*]isoindole-1,3,5,7(2*H*,6*H*)-tetraone (49)

A sealable vial was charged with dibromopyromellitic dianhydride (779 mg, 2.07 mmol, 1.00 equiv.) and 2,4,6-trimethylaniline (560 mg, 582 μL, 4.14 mmol, 2.00 equiv.). Subsequently, acetic acid (7.79 mL) was added, and the reaction mixture was stirred at 100 °C for 4 h. After cooling to room temperature, the mixture was concentrated under reduced pressure. The residue was then dissolved in dichloromethane and the solution was washed with NaHCO$_3$ (3×) and brine. After drying over Na$_2$SO$_4$, the solution was concentrated under reduced pressure, and the crude product was filtrated over a silica plug with dichloromethane. The product (1.20 g, 1.96 mmol, 95% yield) was obtained as a yellow-colorless solid.

^{1}H NMR (400 MHz, CDCl$_3$) δ = 7.02 (s, 4H, 3'-H), 2.35 (s, 6H, C*H*$_3$), 2.13 (s, 12H, C*H*$_3$) ppm.

^{13}C NMR (101 MHz, CDCl$_3$) δ = 162.47 (4C, C-1), 140.17 (4C, C-2), 136.42 (2C, C-1'), 136.17 (2C, C-4'), 129.63 (4C, C-3'), 126.31 (4C, C-2'), 115.41 (2C, C-3), 21.31 (2C, C*H*$_3$), 18.14 (4C, C*H*$_3$) ppm.

MS (FAB, 3-NBA), m/z (%): 610 (9) [M+H]$^+$, 609 (12) [M]$^+$.

HRMS–FAB *(m/z)*: Calc. for [C$_{28}$H$_{23}$O$_4$N$_2$^{79}Br$_2$]$^+$: 609.0019; found: 609.0020.

IR (ATR, ṽ) = 2922 (vw), 1782 (vw), 1715 (vs), 1482 (w), 1443 (w), 1377 (vs), 1353 (w), 1343 (w), 1300 (w), 1133 (vs), 861 (m), 834 (vs), 755 (w), 730 (m), 715 (w), 673 (m), 431 (m) cm^{-1}.

R$_f$ = 0.48 (Cyclohexane : ethyl acetate = 4:1).

4,8-bis(4-bromophenyl)-2,6-dimesitylpyrrolo[3,4-*f*]isoindole-1,3,5,7(2*H*,6*H*)-tetraone (51)

A sealable vial was charged with 4,8-dibromo-2,6-dimesitylpyrrolo[3,4-*f*]isoindole-1,3,5,7(2*H*,6*H*)-tetraone (100 mg, 164 μmol, 1.00 equiv.), (4-(di-phenylamino)phenyl)boronic acid (65.8 mg, 328 μmol, 2.00 equiv.), K$_2$CO$_3$ (104 mg, 753 μmol, 4.59 equiv.) and Pd(PPh$_3$)$_4$ (50.0 mg, 43.3 μmol, 0.26 equiv.) are added. Then, toluene (4.00 mL) and water (1.00 mL) were added, and the solution was stirred at 110 °C for 27 h. Subsequently, the product mixture was poured into an excess of water and extracted with dichloromethane (3×). The combined organic fractions were washed with brine, dried over Na$_2$SO$_4$, and concentrated under reduced pressure. The crude mixture was further purified *via* column chromatography (cyclohexane : dichloromethane = 1:1). The product (74.0 mg, 97.1 μmol, 59% yield) was obtained as a yellow solid.

¹H NMR (400 MHz, CDCl$_3$) δ = 7.63–7.58 (m, 4H, 2''-H), 7.44–7.40 (m, 4H, 3''-H), 6.92 (s, 4H, 3'-H), 2.28 (s, 6H, C*H*$_3$), 2.06 (s, 12H, C*H*$_3$) ppm.

¹³C NMR (101 MHz, CDCl$_3$) δ = 164.25 (4C, C=O), 139.75 (2C, C-4'), 137.79 (2C, C-1''), 135.96 (2C, C-1'), 134.50 (4C, C-1), 131.59 (4C, C-3''), 131.10 (4C, C-2''), 129.34 (4C, C-3'), 129.29 (2C, C-2), 126.48 (4C, C-2'), 124.31 (2C, C-4''), 21.20 (2C, *C*H$_3$), 18.16 (4C, *C*H$_3$) ppm.

MS (FAB, 3-NBA), m/z (%): 764 (4) [M+H]$^+$, 763 (6) [M]$^+$.

HRMS–FAB *(m/z)*: Calc. for [C$_{40}$H$_{31}$O$_4$N$_2$^{79}Br^{81}Br]$^+$: 763.0625, found: 763.0622.

IR (ATR, ṽ) = 2922 (w), 2857 (w), 1765 (w), 1718 (vs), 1589 (w), 1485 (m), 1465 (m), 1402 (w), 1368 (vs), 1300 (m), 1135 (vs), 1064 (m), 1010 (m), 873 (s), 837 (vs), 822 (s), 769 (s), 507 (s) cm^{-1}.

R$_f$ = 0.60 (Cyclohexane : ethyl acetate = 4:1).

184

(4-phenoxazin-10-ylphenyl)boronic acid (52a)

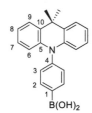

To a 0.1M solution of 10-(4-bromophenyl)phenoxazine (1.50 g, 4.44 mmol, 1.00 equiv.) in absolute THF, n-butyllithium solution (341 mg, 2.13 mL, 5.32 mmol, 2.5 M in hexanes, 1.20 equiv.) was added dropwise while stirring at -78 °C. After 30 min, trimethyl borate (691 mg, 742 µL, 6.65 mmol, 1.50 equiv.) was added dropwise. The mixture was stirred at room temperature for 26 h. Subsequently, the reaction was quenched and extracted with dichloromethane (3×). The combined organic fractions were washed with brine. The crude mixture was then purified via column chromatography (dichloromethane : methanol = 100:1) to afford the product (1.12 g, 3.68 mmol, 83% yield) as a yellow solid.

^{1}H NMR (400 MHz, CDCl$_3$) δ = 8.54–8.46 (m, 2H, 2-H), 7.58–7.50 (m, 2H, 3-H), 6.76–6.57 (m, 6H, 6-H, 7-H, 8-H), 5.99 (d, ^{3}J = 7.6 Hz, 2H, 9-H) ppm.

^{13}C NMR (101 MHz, CDCl$_3$) δ = 144.16 (2C, C-10), 143.91 (C-4), 138.59 (2C, C-2), 136.43 (C-1), 134.18 (2C, C-5), 130.68 (2C, C-3), 123.42 (2C, C-8), 121.75 (2C C-7), 115.76 (2C, C-6), 113.42 (2C, C-9) ppm.

R_f = 0.82 (Cyclohexane : ethyl acetate = 2:1 + 1% NEt$_3$).

(4-(9,9-dimethylacridin-10(9H)-yl)phenyl)boronic acid (52b)

The following procedure was adopted from a literature procedure.[150-151] To a solution of 10-(4-bromophenyl)-9,9-dimethylacridine (238 mg, 653 µmol, 1.00 equiv.) in THF (10.0 mL, abs.), n-butyllithium solution (50.2 mg, 314 µL, 784 µmol, 2.50 M in hexane, 1.20 equiv.) was added dropwise while stirring at -78 °C. After 30 min, trimethyl borate (102 mg, 109 µL, 980 µmol, 1.50 equiv.) was added dropwise. The mixture was stirred at room temperature for 12 h. Subsequently, the reaction was quenched and extracted with dichloromethane (3×). The combined organic fractions were washed with brine and concentrated under reduced pressure. The crude product was then purified via column chromatography (dichloromethane : methanol = 100:1) to afford the product (100 mg, 228 µmol, 35% yield) as a yellow-colorless solid.

¹H NMR (400 MHz, CDCl₃) δ = 8.56 (d, ^3J = 8.2 Hz, 2H, 2-H), 7.54 (d, ^3J = 8.2 Hz, 2H, 3-H), 7.49 (dd, ^3J = 7.4 Hz, ^4J = 1.9 Hz, 2H, 9-H), 7.03–6.91 (m, 4H, 7-H, 8-H), 6.33 (dd, ^3J = 7.9 Hz, ^4J = 1.5 Hz, 2H, 6-H), 1.72 (s, 6H, CH_3) ppm.

IR (ATR, ṽ) = 2958 (w), 2928 (w), 2856 (w), 1591 (vs), 1476 (s), 1463 (s), 1442 (s), 1401 (s), 1332 (vs), 1322 (vs), 1279 (vs), 1266 (vs), 1164 (m), 1101 (s), 1045 (s), 1017 (s), 926 (m), 747 (vs), 738 (vs), 720 (s), 703 (vs), 666 (s), 654 (s), 622 (s), 550 (m), 534 (m), 511 (m), 407 (m) cm⁻¹.

R$_f$ = 0.11 (Cyclohexane : ethyl acetate = 2:1 + NEt₃).

10-(4-(4,4,5,5-tetramethyl-1,3,2-dioxaborolan-2-yl)phenyl)-10H-phenoxazine (52c)

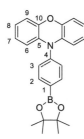

A sealable vial was charged with 10-(4-bromophenyl)phenoxazine (500 mg, 1.48 mmol, 1.00 equiv.), Pd(dppf)Cl₂ (108 mg, 148 µmol, 0.10 equiv.), bis(pinacolato)diboron (451 mg, 1.77 mmol, 1.20 equiv.) and potassium acetate (435 mg, 4.44 mmol, 3.00 equiv.). Subsequently, the vial was evacuated and backfilled with argon three times. Then, 1,4-dioxane (5.00 mL, abs.) was added and the reaction mixture was stirred at 110 °C for 4 h. After cooling to room temperature, the mixture was pulled over a Celite® plug with dichloromethane and the crude product was further purified via column chromatography (cyclohexane : dichloromethane = 4:1). The product (329 mg, 854 µmol, 58% yield) was obtained as a yellow crystalline solid.

¹H NMR (400 MHz, CDCl₃) δ = 8.05–8.01 (m, 2H, 2-H), 7.39–7.31 (m, 2H, 3-H), 6.68 (dd, ^3J = 7.7 Hz, ^4J = 1.8 Hz, 2H, 6-H), 6.63 (td, ^3J = 7.7 Hz, ^4J = 1.6 Hz, 2H, 7-H), 6.56 (td, ^3J = 7.7 Hz, ^4J = 1.8 Hz, 2H, 8-H), 5.91 (dd, ^3J = 7.7 Hz, ^4J = 1.6 Hz, 2H, 9-H), 1.38 (s, 12H, CH_3) ppm.

¹³C NMR (101 MHz, CDCl₃) δ = 144.07 (2C, C-5), 141.84 (C-4), 137.62 (2C, C-2), 134.35 (2C, C-10), 130.22 (2C, C-3), 123.35 (2C, C-8), 121.45 (2C, C-7), 115.55 (2C, C-6), 113.44 (2C, C-9), 84.28 (2C, C$_q$), 25.06 (4C, CH_3) ppm.

¹¹B NMR (128 MHz, CDCl₃) δ = 31.09 ppm.

MS (FAB, 3-NBA), m/z (%): 386 (36) [M+H]⁺, 385 (100) [M]⁺.

HRMS–FAB *(m/z)*: Calc. for $[C_{24}H_{24}O_3N^{11}B]^+$: 385.1844; found: 385.1846.

IR (ATR, ṽ) = 3065 (vw), 2976 (w), 2921 (w), 1599 (m), 1480 (vs), 1462 (m), 1397 (m), 1356 (vs), 1324 (vs), 1290 (s), 1268 (vs), 1262 (vs), 1207 (s), 1139 (vs), 1085 (vs), 1041 (m), 1021 (m), 962 (m), 922 (m), 856 (s), 744 (vs), 654 (vs), 613 (m), 581 (m) cm⁻¹.

R$_f$ = 0.78 (Cyclohexane : ethyl acetate = 4:1).

10-(4-(tributylstannyl)phenyl)-10*H*-phenoxazine (52d)

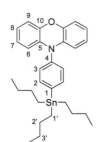

A sealable vial was charged with 10-(4-bromophenyl)phenoxazine (500 mg, 1.48 mmol, 1.00 equiv.), Pd(OAc)₂ (6.64 mg, 29.6 μmol, 0.02 equiv.) and PCy₃ (16.6 mg, 59.1 μmol, 0.04 equiv.). Subsequently, it was evacuated and backfilled with argon three times. Then, toluene (5.00 mL, abs.) was added, and tributyltin (943 mg, 822 μL, 1.63 mmol, 1.10 equiv.) was added while stirring at room temperature. The reaction mixture was subsequently stirred at 110 °C for 17 h. The product mixture was cooled to room temperature and pulled over a Celite® plug with dichloromethane. The crude product was further purified employing column chromatography (cyclohexane : dichloromethane = 100:1 → 50:1). The product (667 mg, 1.22 mmol, 82% yield) was obtained as a colorless oil.

¹H NMR (400 MHz, CDCl₃) δ = 7.66 (d, ^3J = 7.5 Hz, 2H, 2-H), 7.27 (d, ^3J = 7.5 Hz, 2H, 3-H), 6.67 (dd, ^3J = 7.7 Hz, ^4J = 1.9 Hz, 2H, 6-H), 6.65–6.55 (m, 4H, 7-H, 8-H), 5.92 (dd, ^3J = 7.7 Hz, ^4J = 1.8 Hz, 2H, 9-H), 1.70–1.48 (m, 6H, 2'-H), 1.42–1.31 (m, 6H, 3'-H), 1.21–1.02 (m, 6H, 1'-H), 0.92 (t, ^3J = 7.2 Hz, 9H, C*H*₃) ppm.

¹³C NMR (101 MHz, CDCl₃) δ = 144.09 (2C, C-5), 143.24 (C-1), 139.05 (2C, C-2), 138.72 (C-4), 134.68 (2C, C-10), 130.12 (2C, C-3), 123.33 (2C, C-8), 121.26 (2C, C-7), 115.46 (2C, C-6), 113.42 (2C, C-9), 29.24 (3C, C-2'), 27.51 (3C, C-3'), 13.84 (3C, C*H*₃), 9.87 (3C, C-1') ppm.

MS (FAB, 3-NBA), m/z (%): 550 (41) [M+H]⁺, 549 (100) [M]⁺.

HRMS–FAB *(m/z)*: Calc. for $[C_{30}H_{39}ON^{120}Sn]^+$: 549.2048; found: 549.2047.

R_f = 0.73 (Cyclohexane : ethyl acetate = 4:1).

10-(4-((trimethylsilyl)ethynyl)phenyl)-10*H*-phenoxazine (52e)

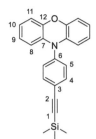

A sealable vial was charged with 10-(4-bromophenyl)phenoxazine (1.00 g, 2.96 mmol, 1.00 equiv.), Pd₂(dba)₃ (271 mg, 296 μmol, 0.10 equiv.), [(*t*Bu)₃PH]BF₄ (257 mg, 887 μmol, 0.30 equiv.) and CuI (84.5 mg, 444 μmol, 0.150 equiv.). It was then evacuated and backfilled with argon three times. Subsequently, NEt₃ (5.00 mL, abs.) and THF (5.00 mL, abs.) were added, and the reaction mixture was degassed. Then, trimethylsilylacetylene (2.90 g, 4.21 mL, 29.6 mmol, 10.0 equiv.) was added dropwise, and the reaction mixture was stirred at 80 °C for 16 h. After cooling to room temperature, the mixture was filtered off, and the residue was washed with dichloromethane. All volatile components were removed under reduced pressure. Subsequently, the crude mixture was pulled over a neutral Al₂O₃ plug with *n*-hexane. After the filtrate was concentrated under reduced pressure, the product (421 mg, 1.18 mmol, 40% yield) was recrystallized from an ethanol solution in the freezer as orange crystals.

¹H NMR (400 MHz, CDCl₃) δ = 7.74–7.62 (m, 2H, 4-H), 7.31–7.27 (m, 2H, 5-H), 6.72–6.62 (m, 4H, 8-H, 9-H), 6.58 (td, ^3J = 7.6 Hz, ^4J = 1.8 Hz, 2H, 10-H), 5.90 (dd, ^3J = 7.9 Hz, ^4J = 1.5 Hz, 2H, 11-H), 0.28 (s, 9H, C*H*₃) ppm.

¹³C NMR (101 MHz, CDCl₃) δ = 144.06 (2C, C-7), 139.20 (C-6), 134.80 (2C, C-4), 134.17 (2C, C-12), 130.97 (2C, C-5), 123.60 (C-3), 123.39 (2C, C-10), 121.66 (2C, C-9), 115.67 (2C, C-8), 113.35 (2C, C-11), 104.12 (C-2), 96.05 (C-1), 0.05 (3C, C*H*₃) ppm.

MS (FAB, 3-NBA), m/z (%): 356 (39) [M+H]⁺, 355 (100) [M]⁺.

HRMS–FAB *(m/z)*: Calc. for $[C_{27}H_{15}O]^+$: 355.1117; found: 355.1116.

188

IR (ATR, ṽ) = 3064 (vw), 2962 (vw), 2156 (w), 1628 (vw), 1592 (w), 1486 (vs), 1462 (m), 1424 (vw), 1402 (w), 1334 (s), 1326 (s), 1292 (m), 1272 (vs), 1247 (s), 1217 (w), 1204 (m), 1154 (w), 1118 (w), 1098 (w), 1044 (w), 1018 (w), 925 (w), 914 (w), 857 (vs), 839 (vs), 826 (vs), 783 (w), 762 (m), 735 (vs), 713 (s), 703 (s), 656 (m), 646 (m), 628 (m), 616 (m), 585 (s), 545 (w), 537 (m), 477 (w), 453 (w), 448 (w), 438 (w), 411 (w) cm^{-1}.

EA: Calc. for $C_{23}H_{21}NOSi$: C 77.71; H 5.95; N 3.94. Found: C 76.81; H 6.25; N 3.80%.

R$_f$ = 0.94 (Cyclohexane : ethyl acetate = 4:1).

10-(4-ethynylphenyl)-10*H*-phenoxazine (52f)

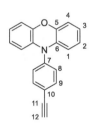

To a solution of 10-(4-((trimethylsilyl)ethynyl)phenyl)-10*H*-phenoxazine (207 mg, 437 μmol, 1.00 equiv.) in THF (2.00 mL), tetrabutylammonium fluoride (228 mg, 873 μL, 873 μmol, 1.00 M in THF, 2.00 equiv.) was added dropwise. The reaction mixture was stirred at room temperature for 19 h. The reaction was stopped by evaporating all volatile components under reduced pressure. The crude product was purified employing column chromatography (cyclohexane : ethyl acetate = 50:1). The product (105 mg, 371 μmol, 85% yield) was obtained as yellow crystals.

^{1}H NMR (400 MHz, CDCl$_3$) δ = 7.75–7.68 (m, 2H, 9-H), 7.35–7.29 (m, 2H, 8-H), 6.70 (dd, ^3J = 7.7 Hz, ^4J = 1.9 Hz, 2H, 4-H), 6.66 (td, ^3J = 7.7 Hz, ^4J = 1.6 Hz, 2H, 3-H), 6.60 (td, ^3J = 7.7 Hz, ^4J = 1.9 Hz, 2H, 2-H), 5.92 (dd, ^3J = 7.7 Hz, ^4J = 1.6 Hz, 2H, 1-H), 3.18 (s, 1H, 12-H) ppm.

^{13}C NMR (101 MHz, CDCl$_3$) δ = 144.06 (2C, C-5), 139.59 (C-7), 134.99 (2C, C-9), 134.12 (2C, C-6), 131.11 (2C, C-8), 123.41 (2C, C-2), 122.58 (C-10), 121.72 (2C, C-3), 115.71 (2C, C-4), 113.36 (2C, C-1), 82.84 (C-11), 78.68 (C-12) ppm.

MS (FAB, 3-NBA), m/z (%): 284 (32) [M+H]$^+$, 283 (100) [M]$^+$.

HRMS–FAB *(m/z)*: [$C_{20}H_{13}ON$]$^+$: 283.0992; found: 283.0990.

IR (ATR, ṽ) = 3302 (w), 3054 (w), 1594 (w), 1482 (vs), 1462 (s), 1402 (w), 1329 (vs), 1289 (m), 1272 (vs), 1256 (s), 1204 (m), 1156 (w), 1119 (w), 1092 (m), 1043 (w), 1018 (w), 918 (m), 867 (m), 827 (m), 738 (vs), 657 (vs), 645 (m), 629 (vs), 613 (s), 585 (s), 533 (w), 514 (m), 459 (w), 452 (w), 426 (w), 394 (m) cm^{-1}.

R_f = 0.74 (Cyclohexane : ethyl acetate = 4:1).

10-(4-bromophenyl)-10H-phenoxazine (55a)

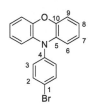

The following procedure was adopted from a literature procedure.[10] Two sealable vials were each charged with 1-bromo-4-iodo-benzene (5.00 g, 17.7 mmol, 1.00 equiv), 10H-phenoxazine (3.24 g, 17.7 mmol, 1.00 equiv), NaOtBu (3.00 g, 31.2 mmol, 1.77 equiv) and CuI (337 mg, 1.77 mmol, 0.10 equiv). The vials were evacuated and backfilled with argon two times. Then, 1,4-dioxane (34.00 mL, abs.) and (1R,2R)-(-)-1,2-cyclohexanediamine (1.01 g, 1.06 mL, 8.84 mmol, 0.50 equiv) were added dropwise to each vial. The reaction mixtures were stirred at 100 °C for 19 h. Then, the product mixture was concentrated under reduced pressure and subsequently extracted with CH$_2$Cl$_2$ (3×). The combined organic fractions were washed with brine, dried over Na$_2$SO$_4$, and concentrated under reduced pressure. The crude residue was purified *via* column chromatography (cyclohexane : dichloromethane = 200:1 → 100:1 → 50:1 → 20:1 → 10:1 → 5:1). The product 10-(4-bromophenyl)phenoxazine (5.16 g, 15.2 mmol, 43% yield) was obtained as a colorless solid.

^{1}H NMR (400 MHz, CDCl$_3$) δ = 7.76–7.69 (m, 2H, 2-H), 7.25–7.20 (m, 2H, 3-H), 6.72–6.63 (m, 4H, 6-H, 7-H), 6.60 (td, ^3J = 7.5 Hz, ^4J = 2.0 Hz, 2H, 8-H), 5.91 (d, ^3J = 7.8 Hz, 2H, 9-H) ppm.

^{13}C NMR (101 MHz, CDCl$_3$) δ = 144.03 (C-4), 138.22(2C, C-10), 134.55 (2C, C-2), 134.10 (2C, C-5), 132.86 (2C, C-3), 123.40 (2C, C-8), 122.48 (C-1), 121.74 (2C, C-7), 115.70 (2C, C-6), 113.29 (2C, C-9) ppm.

MS (FAB, 3-NBA), m/z (%): 339 (20) [M(^{81}Br)]$^+$, 337 (19) [M(^{79}Br)]$^+$.

HRMS–FAB *(m/z)*: Calc. for [C$_{18}$H$_{12}$ON^{79}Br]$^+$: 337.0097, found: 337.0096.

IR (ATR, ṽ) = 3077 (w), 3053 (w), 3029 (w), 2614 (w), 1626 (w), 1589 (w), 1479 (vs), 1462 (vs), 1390 (m), 1330 (vs), 1317 (s), 1289 (s), 1271 (vs), 1207 (s), 1092 (s), 1067 (m), 1009 (s), 817 (m), 793 (m), 731 (vs), 715 (vs), 613 (s), 578 (s), 504 (s), 411 (m), 384 (s) cm^{-1}.

R$_f$ = 0.82 (Cyclohexane : ethyl acetate = 4:1).

10-(4-bromophenyl)-9,9-dimethyl-9,10-dihydroacridine (55b)

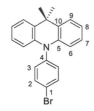

The following procedure was adopted from a literature procedure.[10] Two sealable vials were charged with 1-bromo-4-iodo-benzene (2.50 g, 8.84 mmol, 1.00 equiv.), 9,9-dimethyl-10H-acridine (1.66 g, 7.95 mmol, 0.90 equiv.), NaOtBu (3.40 g, 35.3 mmol, 4.00 equiv.) and CuI (168 mg, 884 µmol, 0.10 equiv.). After evacuating and backfilling with argon three times, 1,4-dioxane (16.00 mL, abs.) was added and (1R,2R)-(−)-1,2-cyclohexanediamine (DACH) (505 mg, 531 µL, 4.42 mmol, 0.50 equiv.) was added dropwise. Subsequently, the reaction mixtures were stirred at 100 °C for 6 h. The product mixtures were then concentrated under reduced pressure and extracted with dichloromethane (3×). The combined organic fractions were extracted with brine, dried over Na$_2$SO$_4$, and concentrated under reduced pressure. The crude mixture was purified *via* column chromatography (cyclohexane : ethyl acetate = 30:1). The product (1.59 g, 4.38 mmol, 50% yield) was isolated as a colorless solid.

^{1}H NMR (400 MHz, CDCl$_3$) δ = 7.79–7.72 (m, 2H, 2-H), 7.46 (dd, ^3J = 7.8 Hz, ^4J = 1.8 Hz, 2H, 7-H), 7.25–7.20 (m, 2H, 3-H), 7.02–6.89 (m, 4H, 6-H, 9-H), 6.25 (dd, ^3J = 7.8 Hz, ^4J = 1.5 Hz, 2H, 8-H), 1.68 (s, 6H, CH$_3$) ppm.

^{13}C NMR (101 MHz, CDCl$_3$) δ = 140.74 (C-4), 140.46 (2C, C-5), 134.36 (2C, C-2), 133.37 (2C, C-3), 130.24 (2C, C-10), 126.55 (2C, C-6), 125.46 (2C, C-7), 122.24 (C-1), 120.95 (2C, C-9), 114.04 (2C, C-8), 36.11 (C-11), 31.38 (2C, CH$_3$) ppm.

MS (FAB, 3-NBA), m/z (%): 365 (18) [M(^{81}Br)]$^+$, 363 (17) [M(^{79}Br)]$^+$.

HRMS–FAB *(m/z)*: Calc. for [C$_{21}$H$_{18}$N^{79}Br]$^+$: 363.0617, found: 363.0616.

IR (ATR, ṽ) = 3075 (w), 3060 (w), 3051 (w), 3026 (w), 2970 (w), 1584 (m), 1468 (s), 1442 (vs), 1319 (vs), 1266 (vs), 1069 (m), 1043 (m), 1013 (s), 918 (m), 744 (vs), 718 (s), 619 (m), 511 (s), 441 (m) cm^{-1}.

R_f = 0.86 (Cyclohexane : ethyl acetate = 4:1).

4,8-bis((4-(10H-phenoxazin-10-yl)phenyl)ethynyl)-2,6-dimesitylpyrrolo[3,4-f]isoindole-1,3,5,7(2H,6H)-tetraone (57)

A sealable vial was charged with 4,8-dibromo-2,6-dimesitylpyrrolo[3,4-f] isoindole-1,3,5,7(2H,6H)-tetraone (200 mg, 328 µmol, 1.00 equiv.), 10-(4-ethynylphenyl)-10H-phenoxazine (186 mg, 655 µmol, 2.00 equiv.), Pd$_2$(dba)$_3$*CHCl$_3$ (13.8 mg, 13.1 µmol, 0.04 equiv.), CuI (2.50 mg, 13.1 µmol, 0.04 equiv.) and triphenylphosphane (8.60 mg, 7.81 µL, 32.8 µmol, 0.10 equiv.). It was then evacuated and backfilled with argon three times. Then, NEt$_3$ (5.00 mL, abs.) and THF (5.00 mL, abs.) were added. Subsequently, the reaction mixture was stirred at 80 °C in the dark for 21 h. After cooling to room temperature, the product mixture was pulled over a Celite® plug with dichloromethane. The filtrate was then poured into an excess of water and extracted with dichloromethane, washed with brine and dried over Na$_2$SO$_4$. After concentration under reduced pressure, the crude product was purified employing column chromatography (cyclohexane : ethyl acetate = 20:1), followed by re-crystallization from a cyclohexane/ethyl acetate mixture and washing with methanol. The product (213 mg, 210 µmol, 64% yield) was obtained as a dark violet solid.

^{1}H NMR (400 MHz, CDCl$_3$) δ = 8.08–7.93 (m, 4H, 4"-H), 7.42–7.33 (m, 4H, 5"-H), 7.05 (s, 4H, 3'-H), 6.70 (dd, ^3J = 7.5 Hz, ^4J = 1.9 Hz, 4H, 2'''-H), 6.66 (td, ^3J = 7.5 Hz, ^4J = 1.5 Hz, 4H, 3'''-H), 6.59 (ddd, ^3J = 7.9 Hz, ^3J = 7.5 Hz, ^4J = 1.9 Hz, 4H, 4'''-H), 5.93 (dd, ^3J = 7.9 Hz, ^4J = 1.5 Hz, 4H, 5'''-H), 2.36 (s, 6H, CH_3), 2.21 (s, 12H, CH_3) ppm.

¹³C NMR (101 MHz, CDCl$_3$) δ = 163.93 (4C, C=O), 144.07 (4C, C-1'''), 140.99 (2C, C-6''), 140.11 (2C, C-4'), 136.57 (4C, C-2'), 136.38 (4C, C-2), 135.74 (4C, C-4''), 133.94 (4C, C-6'''), 131.29 (4C, C-5''), 129.65 (4C, C-3'), 126.52 (2C, C-1'), 123.42 (4C, C-4'''), 122.14 (2C, C-3''), 121.85 (4C, C-3'''), 115.99 (2C, C-3), 115.77 (4C, C-2'''), 113.38 (4C, C-5'''), 105.18 (2C, C-2''), 82.83 (2C, C-1''), 21.32 (2C, CH$_3$), 18.22 (4C, CH$_3$) ppm.

MS (FAB, 3-NBA), m/z (%): 1014 (8) [M]$^+$.

HRMS–FAB *(m/z)*: Calc. for [C$_{68}$H$_{46}$O$_6$N$_4$]$^+$: 1014.3412; found: 1014.3413.

IR (ATR, ṽ) = 2925 (w), 2854 (w), 2210 (w) (C≡C), 1720 (vs) (C=O), 1594 (w), 1514 (m), 1486 (vs), 1463 (m), 1361 (s), 1333 (s), 1302 (m), 1292 (m), 1272 (vs), 1239 (m), 1115 (s), 1040 (m), 867 (m), 830 (vs), 766 (w), 752 (s), 741 (vs), 713 (m), 585 (m), 534 (w) cm^{-1}.

R$_f$ = 0.37 (Cyclohexane : ethyl acetate = 10:1).

5.2.3. **1*H*-Pyrrolo[3,4-*b*]quinoxaline-1,3(2*H*)-dione (PQD)-Based Molecule Design**

Dimethyl 6-bromoquinoxaline-2,3-dicarboxylate (62a)

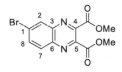

This synthesis was carried out according to a literature procedure.[156] To a mixture of 4-bromobenzene-1,2-diamine (1.20 g, 6.42 mmol, 1.00 equiv.) and dimethyl but-2-ynedioate (913 mg, 790 µL, 6.42 mmol, 1.00 equiv.) in absolute *N*,*N*-dimethylformamide (25.00 mL), PhI(OAc)$_2$ (6.26 g, 19.4 mmol, 3.03 equiv.) was added under argon atmosphere at –20 °C. The mixture was stirred in the dark for 15 h and quenched with aqueous Na$_2$S$_2$O$_3$. The resulting mixture was extracted with dichloromethane (3×). The combined organic extracts were washed with aqueous NaHCO$_3$ solution and brine. Then, they were dried over Na$_2$SO$_4$ and concentrated under reduced pressure. The crude mixture was further purified employing column chromatography (cyclohexane : ethyl acetate = 100:1 → 5:1). The product dimethyl (1.17 g, 3.61 mmol, 56% yield) was obtained as an orange oil that slowly solidified to a pale yellow-orange solid.

^{1}H NMR (400 MHz, CDCl$_3$) δ = 8.44 (d, ^4J = 2.0 Hz, 1H, 2-H), 8.12 (dd, ^3J = 9.0 Hz, ^5J = 0.5 Hz, 1H, 7-H), 8.01 (dd, ^3J = 9.0 Hz, ^4J = 2.1 Hz, 1H, 8-H), 4.09 (s, 3H, C*H*$_3$), 4.09 (s, 3H, C*H*$_3$) ppm.

^{13}C NMR (101 MHz, CDCl$_3$) δ = 165.02 (C=O), 164.94 (C=O), 145.01 (C-4), 143.86 (C-5), 142.02 (C-3), 140.30 (C-6), 136.51 (C-8), 132.25 (C-2), 131.18 (C-7), 127.54 (C-1), 53.84 (2C, C*H*$_3$) ppm.

MS (FAB, 3-NBA), m/z (%): 327 (99) [M(^{81}Br)+H]$^+$, 325 (100) [M(^{79}Br)+H]$^+$.

HRMS–FAB *(m/z)*: Calc. for [C$_{12}$H$_{10}$O$_4$N$_2$^{79}Br]$^+$: 324.9818; found: 324.9819.

R$_f$ = 0.37 (Cyclohexane : ethyl acetate = 4:1).

Dimethyl 6-cyanoquinoxaline-2,3-dicarboxylate (62b)

The following procedure was adopted from a literature procedure.[156] A sealable vial was charged with dimethyl 6-bromo-quinoxaline-2,3-dicarboxylate (500 mg, 1.54 mmol, 1.00 equiv.) and CuCN (344 mg, 3.84 mmol, 2.50 equiv.). It was then evacuated and backfilled with argon. Subsequently, *N,N*-dimethylformamide (8.00 mL, abs.) was added and the reaction was stirred at 140 °C for 24 h. Then, the mixture was poured into an excess of ethyl acetate and washed with 1 M NH_3 until the aqueous phase did not show a blue color. The organic extracts were washed with brine and dried over Na_2SO_4. After concentration under reduced pressure, the crude mixture was further purified employing column chromatography (cyclohexane : ethyl acetate = 20:1 → 8:1). The product (236 mg, 731 µmol, 48% yield) was obtained as a yellow solid.

^{1}H NMR (400 MHz, CDCl$_3$) δ = 8.62 (dd, ^4J = 1.8 Hz, ^5J = 0.6 Hz, 1H, 2-H), 8.36 (dd, ^3J = 8.7 Hz, ^5J = 0.6 Hz, 1H, 7-H), 8.07 (dd, ^3J = 8.7 Hz, ^4J = 1.8 Hz, 1H, 8-H), 4.11 (s, 6H, C*H*$_3$) ppm.

^{13}C NMR (126 MHz, CDCl$_3$) δ = 164.53 (C=O), 164.46 (C=O), 146.26 (C-5), 145.57 (C-4), 142.73 (C-6), 140.59 (C-3), 135.72 (C-2), 133.26 (C-8), 131.64 (C-7), 117.25 (*C*N), 116.38 (C-1), 54.05 (2C, *C*H$_3$) ppm.

MS (FAB, 3-NBA), m/z (%): 272 (12) [M+H]$^+$.

HRMS–FAB *(m/z)*: Calc. for $[C_{13}H_{10}O_4N_3]^+$: 272.0666; found: 272.0664.

IR (ATR, ṽ) = 3078 (vw), 3037 (vw), 2953 (w), 2922 (w), 2853 (w), 2228 (w), 1731 (vs), 1555 (w), 1490 (w), 1441 (vs), 1400 (w), 1349 (m), 1306 (s), 1285 (s), 1224 (vs), 1190 (m), 1166 (s), 1123 (vs), 1067 (vs), 946 (m), 914 (m), 858 (s), 849 (vs), 823 (m), 813 (s), 778 (s), 701 (m), 619 (s), 567 (w), 544 (w), 526 (w), 499 (m), 487 (m), 450 (w), 435 (m), 377 (vs) cm^{-1}.

R$_f$ = 0.31 (Cyclohexane : ethyl acetate = 2:1).

6-Bromoquinoxaline-2,3-dicarboxylic acid (63)

In a round flask, dimethyl 6-bromoquinoxaline-2,3-dicarboxylate (3.19 g, 9.81 mmol, 1.00 equiv.) was stirred in 1 M NaOH solution (3.92 g, 98.1 mL, 98.10 mmol, 10.00 equiv.) at 110 °C for 4 h. Subsequently, the mixture was acidified with 6 M HCl (16.4 mL, 98.1 mmol, 10.00 equiv.) until it reached pH 2. Then, the product mixture was extracted with EtOAc : MeOH (9:1), dried over Na_2SO_4, and concentrated under reduced pressure. The product (2.62 g, 8.82 mmol, 94% yield) was obtained as a pale red solid.

^{1}H NMR (400 MHz, DMSO-d$_6$) δ = 8.51 (t, ^4J = 1.4 Hz, 1H, 8-H), 8.17 (d, ^4J = 1.3 Hz, 2H, 2-H, 3-H) ppm.

^{13}C NMR (101 MHz, DMSO-d$_6$) δ = 165.84 (C-10), 165.78 (C-9), 146.06 (C-5), 145.11 (C-6), 141.11 (C-4), 139.40 (C-7), 135.78 (C-2), 131.31 (C-8), 131.11 (C-3), 126.01 (C-1) ppm.

MS (FAB, 3-NBA), m/z (%): 299 (48) [M(^{81}Br)+H]$^+$, 297 (51) [M(^{79}Br)+H]$^+$.

HRMS–FAB *(m/z)*: Calc. for [C$_{10}$H$_6$O$_4$N$_2$^{79}Br]$^+$: 296.9505; found: 296.9504.

IR (ATR, ṽ) = 3074 (w), 2877 (w), 2564 (w), 1704 (vs), 1595 (m), 1555 (w), 1441 (vs), 1354 (w), 1279 (m), 1241 (s), 1218 (vs), 1159 (vs), 1128 (vs), 1067 (vs), 1055 (vs), 924 (vs), 880 (s), 837 (vs), 819 (vs), 776 (vs), 741 (vs), 734 (vs), 596 (s), 460 (m), 419 (vs), 401 (vs) cm^{-1}.

EA: Calc. for C$_{10}$H$_5$BrN$_2$O$_4$: C 40.43; H 1.70; N 9.43. Found: C 40.68; H 1.89; N 9.20%.

6-Bromo-2-mesityl-1*H*-pyrrolo[3,4-*b*]quinoxaline-1,3(2*H*)-dione (64a)

To a flame-dried Schlenk flask, *N,N*-dimethylformamide (80 mL, abs.) was added. Then, 6-bromoquinoxaline-2,3-dicarboxylic acid (2.55 g, 8.58 mmol, 1.00 equiv.) and HATU (6.53 g, 17.2 mmol, 2.00 equiv.) were added in portions. The mixture was stirred at 0 °C for 30 min. In a separate sealable vial, a solution of 2,4,6-trimethylaniline (1.28 g, 1.33 mL, 9.44 mmol, 1.10 equiv.) in *N,N*-dimethylformamide (22.00 mL, abs.) was prepared. To this solution, DIPEA (5.55 g, 7.48 mL, 42.9 mmol,

5.00 equiv.) was added dropwise while stirring at 0 °C. The resulting mixture was stirred at 0 °C. After 30 min, the amine solution was added dropwise to the active ester-containing solution. The reaction mixture was stirred at room temperature for 4 d and subsequently poured into an excess of water and extracted with dichloromethane (3×). The combined organic extracts were washed with brine, dried over Na_2SO_4 and concentrated under reduced pressure. The product (2.22 g, 5.60 mmol, 65% yield) was recrystallized from a cyclohexane/ethyl acetate solution as a pale yellow-orange solid.

¹H NMR (400 MHz, CDCl$_3$) δ = 8.64 (d, ^4J = 2.1 Hz, 1H, 8-H), 8.33 (d, ^3J = 9.0 Hz, 1H, 5-H), 8.13 (dd, ^3J = 9.0 Hz, ^4J = 2.1 Hz, 1H, 6-H), 7.06 (s, 2H, 3'-H), 2.36 (s, 3H, CH_3), 2.17 (s, 6H, CH_3) ppm.

¹³C NMR (101 MHz, CDCl$_3$) δ = 162.80 (C=O), 162.70 (C=O), 145.69 (2C, C-2, C-3), 144.96 (C-8a), 143.37 (C-4a), 140.26 (C-4'), 137.23 (C-6), 135.82 (2C, C-2'), 133.49 (C-8), 132.33 (C-5), 129.59 (2C, C-3'), 128.55 (C-7), 126.17 (C-1'), 21.17 (CH_3), 18.11 (2C, CH_3) ppm.

MS (FAB, 3-NBA), m/z (%): 398 (31) [M(^{81}Br)]$^+$, 397 (18) [M(^{79}Br)+H]$^+$, 396 (31) [M(^{79}Br)]$^+$.

HRMS–FAB *(m/z)*: Calc. for [C$_{19}$H$_{15}$O$_2$N$_3$^{79}Br]$^+$: 396.0342; found: 396.0343.

IR (ATR, ṽ) = 3087 (w), 3067 (w), 2915 (w), 2864 (w), 1790 (m), 1734 (vs), 1565 (m), 1483 (m), 1368 (vs), 1286 (m), 1128 (vs), 1111 (vs), 1103 (vs), 1052 (s), 1035 (s), 932 (s), 861 (vs), 841 (vs), 826 (vs), 745 (s), 623 (s), 568 (vs), 411 (vs) cm^{-1}.

R$_f$ = 0.48 (Cyclohexane : ethyl acetate = 4:1).

6-Bromo-2-(4-methoxyphenyl)-1*H*-pyrrolo[3,4-*b*]quinoxaline-1,3(2*H*)-dione (64b)

A sealable vial is charged with 6-bromoquinoxaline-2,3-dicarboxylic acid (100 mg, 337 µmol, 1.00 equiv.) and HATU (256 mg, 673 µmol, 2.00 equiv.) and evacuated and backfilled with argon. Then, *N*,*N*-dimethylformamide (2.00 mL, abs.) was added and the mixture was stirred at 0 °C for 30 min. A separate vial was evacuated and backfilled with argon. *p*-Anisidine (41.5 mg, 337 µmol, 1.00 equiv.)

and *N*,*N*-dimethylformamide (2.00 mL, abs.) were added and cooled to 0 °C. Then, DIPEA (218 mg, 293 μL, 1.68 mmol, 5.00 equiv.) was added dropwise and the resulting mixture was stirred at 0 °C. After 30 min, the amine solution was added dropwise to the active ester-containing solution. The reaction mixture was stirred at room temperature for 16 h and subsequently poured into an excess of water and extracted with dichloromethane (3×). The combined organic extracts were washed with brine, dried over Na_2SO_4 and concentrated under reduced pressure. The crude mixture was washed with methanol and cyclohexane to give the product (65.0 mg, 169 μmol, 50% yield) as a pale-yellow solid. A ^{13}C NMR could only be obtained in THF-d$_8$ due to solubility issues. The resulting spectrum displayed rotamers which are indicated where needed using the indices a/b.

^{1}H NMR (400 MHz, CDCl$_3$) δ = 8.64 (d, ^4J = 2.2 Hz, 1H, 2-H), 8.32 (d, ^3J = 9.0 Hz, 1H, 7-H), 8.12 (dd, ^3J = 9.0 Hz, ^4J = 2.2 Hz, 1H, 8-H), 7.49–7.42 (m, 2H, 2'-H), 7.11–7.06 (m, 2H, 3'-H), 3.88 (s, 3H, CH_3) ppm.

^{13}C NMR (126 MHz, THF-d$_8$) δ = 164.26 (C$_a$=O), 163.42 (C$_b$=O), 163.32 (C$_b$-4'), 162.30 (C$_a$-4'), 147.94 (C-5), 147.36 (C-4), 145.61 (C-3), 144.01 (C-6), 137.23 (C-8), 134.22 (C-2), 133.35 (C-7), 129.69 (2C, C$_{a/b}$-2'), 129.62 (2C, C$_{a/b}$-2'), 129.10 (C$_{a/b}$-1'), 129.07 (C$_{a/b}$-1'), 127.99 (C-1), 116.82 (2C, C$_{a/b}$-3'), 116.64 (2C, C$_{a/b}$-3'), 79.64 (CH$_3$) ppm.

MS (EI, 70 eV, 150 °C), m/z (%): 385 (100) [M(^{81}Br)+H]$^+$, 383 (98) [M(^{79}Br)+H]$^+$.

HRMS–EI *(m/z)*: Calc. for [C$_{17}$H$_{10}$O$_3$N$_3$^{79}Br]$^+$: 382.9900; found: 382.9899.

IR (ATR, ṽ) = 3075 (w), 3012 (w), 2959 (w), 2836 (w), 1802 (w), 1730 (vs), 1720 (vs), 1609 (m), 1579 (m), 1516 (s), 1490 (s), 1465 (m), 1391 (m), 1248 (s), 1238 (s), 1177 (m), 1143 (vs), 1126 (vs), 1101 (vs), 1054 (m), 1028 (vs), 1010 (m), 941 (m), 929 (m), 888 (s), 836 (vs), 809 (vs), 795 (vs), 738 (vs), 625 (s), 595 (m), 574 (vs), 530 (vs), 401 (vs) cm^{-1}.

EA: Calc. for C$_{17}$H$_{10}$BrN$_3$O$_3$: C 53.15; H 2.62; N 10.94. Found: C 52.89; H 2.46; N 11.00%.

R$_f$ = 0.32 (Cyclohexane : ethyl acetate = 4:1).

6-Bromo-2-(3,4,5-trimethoxyphenyl)-1*H*-pyrrolo[3,4-*b*]quinoxaline-1,3(2*H*)-dione (64c)

A sealable vial was charged with 6-bromoquinoxaline-2,3-dicarboxylic acid (100 mg, 337 µmol, 1.00 equiv.) and HATU (256 mg, 673 µmol, 2.00 equiv.) and evacuated and backfilled with argon. Then, *N*,*N*-dimethylformamide (2 mL, abs.) was added, and the mixture was stirred at 0 °C for 30 min. A separate vial was evacuated and backfilled with argon. 3,4,5-Trimethoxyaniline (61.7 mg, 337 µmol, 1.00 equiv.) and *N*,*N*-dimethylformamide (2.00 mL, abs.) were added, and the mixture was cooled to 0 °C. Then, DIPEA (218 mg, 293 µL, 1.68 mmol, 5.00 equiv.) was added dropwise, and the resulting mixture was stirred at 0 °C. After 30 min, the amine solution was added dropwise to the active ester-containing solution. Subsequently, the reaction mixture was stirred at room temperature for 14 h. Thereafter, it was poured into an excess of water and extracted with dichloromethane (3×). The combined organic extracts were washed with brine, dried over Na_2SO_4, and concentrated under reduced pressure. The resulting crude mixture was washed with methanol and cyclohexane in an ultrasonic bath to obtain the product (137 mg, 308 µmol, 92% yield) as a salmon-colored solid.

^{1}H NMR (400 MHz, CDCl$_3$) δ = 8.65 (d, ^4J = 2.2 Hz, 1H, 2-H), 8.33 (d, ^3J = 9.0 Hz, 1H, 7-H), 8.13 (dd, ^3J = 9.0 Hz, ^4J = 2.2 Hz, 1H, 8-H), 6.73 (s, 2H, 2'-H), 3.91 (s, 3H, C*H*$_3$), 3.91 (s, 6H, C*H*$_3$) ppm.

^{13}C NMR (101 MHz, CDCl$_3$) δ = 162.96 (C=O), 162.85 (C=O), 153.85 (2C, C-3'), 145.45 (C-4), 145.15 (C-3), 144.73 (C-5), 143.56 (C-6), 138.88 (C-4'), 137.45 (C-8), 133.64 (C-2), 132.43 (C-7), 128.83 (C-1), 126.11 (C-1'), 104.39 (2C, C-2'), 61.10 (*C*H$_3$), 56.50 (2C, *C*H$_3$) ppm.

MS (FAB, 3-NBA), m/z (%): 446 (2) [M(^{81}Br)+H]$^+$, 445 (3) [M(^{81}Br)]$^+$, 444 (2) [M(^{79}Br)+H]$^+$, 443 (3) [M(^{79}Br)]$^+$.

HRMS–FAB *(m/z)*: [C$_{19}$H$_{14}$O$_5$N$_3$^{79}Br]$^+$: 443.0111; found: 443.0113.

IR (ATR, ṽ) = 2932 (w), 2837 (w), 1793 (vw), 1721 (vs), 1598 (m), 1568 (w), 1509 (m), 1490 (w), 1418 (m), 1320 (w), 1237 (s), 1129 (vs), 1102 (s), 1051 (m), 1000 (m), 847 (m), 822 (w), 806 (s), 625 (m), 615 (m), 407 (m) cm^{-1}.

R$_f$ = 0.12 (Cyclohexane : ethyl acetate = 4:1).

6-Bromo-2-(4-(diphenylamino)phenyl)-1*H*-pyrrolo[3,4-*b*]quinoxaline-1,3(2*H*)-dione (64d)

A sealable vial was charged with 6-bromoquinoxaline-2,3-dicarboxylic acid (36.5 mg, 123 µmol, 1.00 equiv.) and HATU (93.5 mg, 246 µmol, 2.00 equiv.) and evacuated and backfilled with argon. Then, *N,N*-dimethylformamide (1.46 mL, abs.) was added, and the mixture was stirred at 0 °C for 30 min. A separate vial was charged with (4-aminophenyl)-diphenyl-amine (32.0 mg, 123 µmol, 1.00 equiv.) and subsequently evacuated and backfilled with argon. Then, *N,N*-dimethylformamide (1.46 mL, abs.) was added and the mixture cooled to 0 °C. DIPEA (79.4 mg, 107 µL, 615 µmol, 5.00 equiv.) was added dropwise and the resulting mixture was stirred at 0 °C. After 30 min, the amine solution was added dropwise to the active ester-containing solution. The reaction mixture was stirred at room temperature for 4 h and subsequently poured into an excess of water and extracted with dichloromethane (3×). The combined organic extracts were washed with brine, dried over Na_2SO_4, and concentrated under reduced pressure. The product (56.0 mg, 107 µmol, 87% yield) was obtained after washing the crude mixture with methanol in an ultrasonic bath as an orange powder.

^{1}H NMR (400 MHz, $CDCl_3$) δ = 8.63 (d, ^4J = 2.2 Hz, 1H, 7-H), 8.32 (d, ^3J = 9.0 Hz, 1H, 4-H), 8.11 (dd, ^3J = 9.0 Hz, ^4J = 2.2 Hz, 1H, 5-H), 7.38–7.28 (m, 6H, 2'-H, 7'-H), 7.21–7.15 (m, 6H, 3'-H, 6'-H), 7.13–7.05 (m, 2H, 8'-H) ppm.

^{13}C NMR (101 MHz, $CDCl_3$) δ = 163.08 (C=O), 162.97 (C=O), 148.71 (C-4'), 147.21 (2C, C-5'), 145.63 (C-8), 145.09 (C-1), 144.92 (C-2), 143.49 (C-3), 137.27 (C-5), 133.57 (C-7), 132.38 (C-4), 129.66 (4C, C-7'), 128.61 (C-6), 127.13 (2C, C-3'), 125.43 (4C, C-6'), 124.04 (2C, C-8'), 123.73 (C-1'), 122.49 (2C, C-2') ppm.

MS (FAB, 3-NBA), m/z (%): 523 (17) [M(^{81}Br)+H]$^+$, 522 (24) [M(^{81}Br)]$^+$, 521 (17) [M(^{79}Br)+H]$^+$, 520 (18) [M(^{79}Br)]$^+$.

HRMS–FAB *(m/z)*: [$C_{28}H_{17}O_2N_4{}^{79}Br$]$^+$: 520.0529; found: 520.0527.

IR (ATR, ṽ) = 3063 (w), 1724 (vs), 1587 (s), 1507 (vs), 1489 (vs), 1374 (s), 1316 (s), 1269 (vs), 1147 (s), 1125 (vs), 1094 (s), 847 (s), 812 (s), 749 (vs), 739 (s), 694 (vs), 618 (s), 578 (s), 503 (vs), 408 (s) cm^{-1}.

EA: Calc. for $C_{28}H_{17}BrN_4O_2$: C 64.50; H 3.29; N 10.75. Found: C 64.02; H 3.14; N 10.79%.

R$_f$ = 0.48 (Cyclohexane : ethyl acetate = 4:1).

6-Bromo-2-(4-fluorophenyl)-1*H*-pyrrolo[3,4-*b*]quinoxaline-1,3(2*H*)-dione (64e)

A sealable vial was charged with 6-bromoquinoxaline-2,3-dicarboxylic acid (100 mg, 337 µmol, 1.00 equiv.) and HATU (256 mg, 673 µmol, 2.00 equiv.) and evacuated and backfilled with argon. Then, *N*,*N*-dimethylformamide (2.00 mL, abs.) was added, and the mixture was stirred at 0 °C for 30 min. A separate vial was evacuated and backfilled with argon. To this vial, 4-fluoroaniline (37.4 mg, 31.9 µL, 337 µmol, 1.00 equiv.) and *N*,*N*-dimethylformamide (2.00 mL, abs.) were added and cooled to 0 °C. Then, DIPEA (218 mg, 293 µL, 1.68 mmol, 5.00 equiv.) was added dropwise, and the resulting mixture was stirred at 0 °C. After 30 min, the amine solution was added dropwise to the active ester-containing solution. The reaction mixture was stirred at room temperature for 16 h and was subsequently poured into an excess of water and extracted with dichloromethane (3×). The combined organic extracts were washed with brine, dried over Na_2SO_4, and concentrated under reduced pressure. The crude mixture was washed with methanol and dichloromethane in an ultrasonic bath to obtain the product (45.0 mg, 121 µmol, 36% yield) as a pale pink solid.

^{1}H NMR (400 MHz, DMSO-d₆) δ = 8.74 (d, $^4J_{H,H}$ = 2.2 Hz, 1H, 2-H), 8.38 (d, $^3J_{H,H}$ = 9.0 Hz, 1H, 7-H), 8.29 (dd, $^3J_{H,H}$ = 9.0 Hz, $^4J_{H,H}$ = 2.2 Hz, 1H, 8-H), 7.55 (ddt, $^3J_{H,H}$ = 8.3 Hz, $^3J_{H,F}$ = 5.4 Hz, $^4J_{H,H}$ = 2.7 Hz, 2H, 3'-H), 7.50–7.43 (m, 2H, 2'-H) ppm.

HRMS–FAB *(m/z)*: Calc. for $[C_{16}H_8O_2N_3{}^{79}BrF]^+$: 371.9778; found: 371.9778.

IR (ATR, ṽ) = 3072 (w), 1799 (w), 1734 (vs), 1714 (m), 1604 (w), 1572 (w), 1514 (vs), 1490 (s), 1381 (s), 1231 (s), 1150 (s), 1140 (vs), 1126 (vs), 1106 (s), 1086 (s), 1055 (m), 1014 (w),

943 (w), 933 (w), 888 (m), 850 (vs), 820 (s), 810 (vs), 754 (w), 739 (s), 626 (m), 588 (m), 571 (vs), 541 (w), 524 (s), 510 (s), 487 (w), 409 (vs), 388 (w) cm^{-1}.

R$_f$ = 0.28 (Dichloromethane).

6-Bromo-2-(4-ethynylphenyl)-1*H*-pyrrolo[3,4-*b*]quinoxaline-1,3(2*H*)-dione (64f)

A sealable vial was charged with 6-bromoquinoxaline-2,3-dicarboxylic acid (50.0 mg, 168 μmol, 1.00 equiv.) and HATU (128 mg, 337 μmol, 2.00 equiv.) and evacuated and backfilled with argon. Then, *N,N*-dimethylformamide (1 mL, abs.) was added, and the mixture was stirred at 0 °C for 30 min. A separate vial was evacuated and backfilled with argon. 4-Ethynylaniline (35.5 mg, 303 μmol, 1.80 equiv.) and *N,N*-dimethylformamide (1.00 mL, abs.) were added and cooled to 0 °C. Then, DIPEA (109 mg, 147 μL, 842 μmol, 5.00 equiv.) was added dropwise and the resulting mixture was stirred at 0 °C. After 30 min, the amine solution was added dropwise, to the active ester-containing solution. The reaction mixture was stirred at room temperature for 16 h and subsequently poured into an excess of water and extracted with dichloromethane (3×). The combined organic extracts were washed with brine, dried over Na$_2$SO$_4$, and concentrated under reduced pressure. The crude mixture was washed with methanol in an ultrasonic bath. After separating the solid from the methanol solution, the product (40.0 mg, 106 μmol, 63% yield) was obtained as a pale red solid.

^{1}H NMR (400 MHz, CDCl$_3$) δ = 8.65 (d, ^4J = 2.2 Hz, 1H, 2-H), 8.33 (d, ^3J = 9.0 Hz, 1H, 7-H), 8.14 (dd, ^3J = 9.0 Hz, ^4J = 2.2 Hz, 1H, 8-H), 7.72–7.65 (m, 2H, 3'-H), 7.59–7.52 (m, 2H, 2'-H), 3.18 (s, 1H, 6'-H) ppm.

^{13}C NMR (126 MHz, THF-d$_8$) δ = 163.23 (C=O), 163.12 (C=O), 147.87 (C-4), 147.28 (C-5), 145.64 (C-3), 144.05 (C-6), 137.25 (C-8), 134.23 (C-2), 133.42 (2C, C-3'), 133.35 (C-7), 133.21 (C-1'), 128.02 (C-1), 127.36 (2C, C-2'), 123.59 (C-4'), 83.66 (C-5''), 80.23 (C-6'') ppm.

HRMS–FAB *(m/z)*: Calc. for [C$_{18}$H$_9$O$_2$N$_3$^{79}Br]$^+$: 377.9873; found: 377.9872.

IR (ATR, ṽ) = 3298 (w), 1734 (vs), 1725 (vs), 1679 (w), 1579 (w), 1509 (w), 1492 (m), 1387 (m), 1276 (w), 1150 (vs), 1126 (vs), 1096 (m), 1055 (w), 933 (w), 888 (w), 849 (w), 836 (s), 815 (vs), 738 (s), 650 (m), 637 (m), 623 (vs), 588 (m), 561 (s), 541 (m), 487 (w), 455 (w), 419 (m), 401 (s) cm^{-1}.

R_f = 0.28 (Dichloromethane).

6-Bromo-2-(4-((trimethylsilyl)ethynyl)phenyl)-1*H*-pyrrolo[3,4-*b*]quinoxaline-1,3(2*H*)-dione (64g)

A sealable vial was charged with 6-bromo-quinoxaline-2,3-dicarboxylic acid (500 mg, 1.68 mmol, 1.00 equiv.) and HATU (1.28 g, 3.37 mmol, 2.00 equiv.) and evacuated and backfilled with argon. Then, *N*,*N*-dimethylformamide (5.00 mL, abs.) was added, and the mixture was stirred at 0 °C for 30 min. A separate vial was charged with 4-(2-trimethylsilylethynyl)aniline (351 mg, 1.85 mmol, 1.10 equiv.) and evacuated and backfilled with argon. Then, *N*,*N*-di-methylformamide (3.00 mL, abs.) was added. Subsequently, DIPEA (1.09 g, 1.47 mL, 8.42 mmol, 5.00 equiv.) was added dropwise while stirring at 0 °C. After 30 min, the amine solution was added dropwise to the active ester-containing solution. The reaction mixture was stirred at room temperature for 25 h and subsequently poured into an excess of water and extracted with ethyl acetate (3×). The combined organic extracts were washed with brine, dried over Na$_2$SO$_4$, and concentrated under reduced pressure. The crude mixture was further purified employing column chromatography with cyclohexane:ethyl acetate (40:1 → 8:1). The product (56.0 mg, 124 μmol, 7% yield) was obtained as an orange solid.

^{1}H NMR (400 MHz, CDCl$_3$) δ = 8.65 (d, ^4J = 2.1 Hz, 1H, 2-H), 8.33 (d, ^3J = 8.9 Hz, 1H, 7-H), 8.13 (dd, ^3J = 9.0 Hz, ^4J = 2.2 Hz, 1H, 8-H), 7.65 (d, ^3J = 8.6 Hz, 2H, 2'-H), 7.53 (d, ^3J = 8.6 Hz, 2H, 3'-H), 0.28 (s, 9H, C*H*$_3$) ppm.

R_f = 0.32 (Cyclohexane : ethyl acetate = 4:1).

6-(4-(diphenylamino)phenyl)-2-mesityl-1*H*-pyrrolo[3,4-*b*]quinoxaline-1,3(2*H*)-dione (66a)

A sealable vial was charged with 6-bromo-2-mesityl-1*H*-pyrrolo[3,4-*b*]quinoxaline-1,3(2*H*) dione (100 mg, 252 µmol, 1.00 equiv.), [4-(diphenylamino)phenyl] boronic acid (73.0 mg, 252 µmol, 1.00 equiv.), K_2CO_3 (69.8 mg, 505 µmol, 2.00 equiv.) and Pd(P(Ph)$_3$)$_4$ (58.3 mg, 50.5 µmol, 0.20 equiv.). Then, the vial was evacuated and backfilled with argon. Subsequently, THF (5.00 mL, abs.) was added, and the reaction mixture was reacted in a microwave reactor for 2 h at max. 150 W and cooling. After the reaction was done, all volatile components were removed and the crude mixture was purified employing column chromatography (cyclohexane : ethyl acetate = 20:1). The product (122 mg, 218 µmol, 86% yield) was obtained as a bright red solid.

¹H NMR (400 MHz, CDCl$_3$) δ = 8.60 (d, ^4J = 2.1 Hz, 1H, 7-H), 8.47 (d, ^3J = 8.9 Hz, 1H, 4-H), 8.31 (dd, ^3J = 8.9 Hz, ^4J = 2.1 Hz, 1H, 5-H), 7.73–7.66 (m, 2H, 2"-H), 7.36–7.30 (m, 4H, 7"-H), 7.24–7.17 (m, 6H, 3"-H, 6"-H), 7.15–7.09 (m, 2H, 8"-H), 7.05 (s, 2H, 3'-H), 2.36 (s, 3H, C*H*$_3$), 2.18 (s, 6H, C*H*$_3$) ppm.

¹³C NMR (101 MHz, CDCl$_3$) δ = 163.54 (2C, C-1, C-10), 149.53 (C-4"), 147.16 (2C, C-5"), 145.95 (C-2), 145.59 (C-6), 145.39 (C-9), 144.04 (C-8), 143.89 (C-3), 140.22 (C-1'), 136.09 (2C, C-2'), 132.89 (C-5), 131.61 (C-4), 130.73 (C-1"), 129.70 (4C, C-7"), 129.69 (2C, C-3'), 128.51 (2C, C-2"), 126.81 (C-7), 126.51 (C-4'), 125.51 (4C, C-6"), 124.17 (2C, C-8"), 122.71 (2C, C-3"), 21.30 (CH$_3$), 18.26 (2C, CH$_3$) ppm.

MS (FAB, 3-NBA), m/z (%): 561 (100) [M+H]$^+$, 560 (89) [M]$^+$.

HRMS–FAB *(m/z)*: Calc. for [C$_{37}$H$_{29}$O$_2$N$_4$]$^+$: 561.2285; found: 561.2285.

IR (ATR, ṽ) = 1785 (w) (C=O), 1734 (vs) (C=O), 1592 (m), 1561 (w), 1494 (vs), 1485 (vs), 1443 (m), 1367 (vs), 1322 (s), 1292 (s), 1279 (vs), 1264 (vs), 1239 (s), 1166 (m), 1128 (vs), 1106 (s), 1041 (m), 856 (m), 820 (vs), 752 (s), 698 (vs), 533 (s), 513 (vs), 415 (m), 405 (s) cm^{-1}.

R$_f$ = 0.69 (Cyclohexane : ethyl acetate = 2:1).

6-(4-(diphenylamino)phenyl)-2-(4-methoxyphenyl)-1*H*-pyrrolo[3,4-*b*]quinoxaline-1,3(2*H*)-dione (66b)

A sealable vial was charged with 6-bromo-2-(4-methoxyphenyl)pyrrolo[3,4-*b*] quinoxaline-1,3-dione (30.0 mg, 78.1 μmol, 1.00 equiv.), [4-(diphenylamino)phenyl] boronic acid (22.6 mg, 78.1 μmol, 1.00 equiv.), K_2CO_3 (32.4 mg, 234 μmol, 3.00 equiv.) and Pd(PPh$_3$)$_4$ (9.02 mg, 7.81 μmol, 0.10 equiv.). It was then evacuated and backfilled with argon three times. Then, THF (3.00 mL, abs.) was added and the reaction mixtue was stirred at 70 °C for 24 h. Subsequently, the mixture was pulled over Celite and concentrated under reduced pressure. After recrystallization from methanol, the product (24.0 mg, 43.7 μmol, 56% yield) was obtained as a yellow-orange solid.

¹H NMR (400 MHz, CDCl$_3$) δ = 8.57 (d, ^4J = 2.1 Hz, 1H, 7-H), 8.43 (d, ^3J = 8.9 Hz, 1H, 4-H), 8.33 (dd, ^3J = 8.9 Hz, ^4J = 2.1 Hz, 1H, 5-H), 7.77–7.72 (m, 2H, 2"-H), 7.46–7.40 (m, 2H, 2'-H), 7.37–7.31 (m, 4H, 7"-H), 7.21–7.16 (m, 6H, 3"-H, 6"-H), 7.15–7.07 (m, 4H, 3'-H, 8"-H), 3.89 (s, 3H, C*H*$_3$) ppm.

¹³C NMR (101 MHz, CDCl$_3$) δ = 163.96 (2C, C=O), 160.37 (C-4'), 149.67 (C-4"), 147.50 (2C, C-5"), 146.02 (C-6), 145.77 (C-8), 145.60 (C-2), 144.28 (C-3), 144.13 (C-1), 133.00 (C-5), 131.80 (C-4), 131.16 (C-1"), 129.90 (4C, C-7"), 128.78 (2C, C-2"), 128.30 (2C, C-2'), 127.09 (C-7), 125.71 (4C, C-6"), 124.34 (2C, C-3'), 123.89 (C-1'), 122.89 (2C, C-3"), 114.92 (2C, C-8"), 56.01 (*C*H$_3$) ppm.

MS (FAB, 3-NBA), m/z (%): 549 (16) [M+H]$^+$, 548 (14) [M]$^+$.

HRMS–FAB *(m/z)*: Calc. for [C$_{35}$H$_{25}$O$_3$N$_4$]$^+$: 549.1921; found: 549.1919.

IR (ATR, ṽ) = 1796 (w), 1727 (vs), 1611 (m), 1589 (s), 1575 (m), 1513 (vs), 1489 (vs), 1441 (m), 1387 (s), 1317 (vs), 1299 (vs), 1283 (m), 1248 (vs), 1193 (m), 1179 (m), 1150 (m), 1132 (vs), 1099 (s), 1072 (m), 1037 (s), 888 (m), 824 (vs), 815 (vs), 803 (s), 754 (vs), 741 (s), 696 (vs), 671 (m), 619 (w), 575 (m), 527 (vs), 518 (vs), 499 (s), 489 (m), 414 (m) cm^{-1}.

R_f = 0.54 (Cyclohexane : ethyl acetate = 2:1).

2-mesityl-6-(10H-phenoxazin-10-yl)-1H-pyrrolo[3,4-b]quinoxaline-1,3(2H)-dione (67)

A sealable microwave tube was charged with 6-bromo-2-mesityl-1H-pyrrolo[3,4-b]quinoxaline-1,3 (2H)dione (100 mg, 252 μmol, 1.00 equiv.), 10H-phenoxazine (54.0 mg, 295 μmol, 1.17 equiv.), K$_2$CO$_3$ (70.0 mg, 506 μmol, 2.01 equiv.) and Pd(PPh$_3$)$_4$ (30.0 mg, 26.0 μmol, 0.10 equiv.). It was then sealed and evacuated and backfilled with argon three times. Subsequently, N,N-dimethylformamide (2.00 mL, abs.), and toluene (500 μL, abs.) were added, and the reaction mixture was degassed for two minutes with argon gas. Then, it was transferred to the microwave and reacted under the following conditions: 150 °C, 150 W, stirring on, cooling on, 40 min. The reaction progress was monitored after 15 min using ASAP-MS. After 15 min, the educt was still visible; thus, the reaction was further reacted for 25 min. After the product mixture was cooled to room temperature, the mixture was pulled over a Celite® plug and extracted with water and dichloromethane to remove DMF. Subsequently, the combined organic fractions were concentrated under reduced pressure and the crude product was purified employing column chromatography (cyclohexane : ethyl acetate = 20:1). The product (78.0 mg, 156 μmol, 62% yield) was obtained as a dark violet solid.

^{1}H NMR (400 MHz, CDCl$_3$) δ = 8.58 (d, ^3J = 9.1 Hz, 1H, 7-H), 8.47 (d, ^4J = 2.4 Hz, 1H, 2-H), 8.06 (dd, ^3J = 9.1 Hz, ^4J = 2.4 Hz, 1H, 8-H), 7.06 (s, 2H, 3'-H), 6.93–6.83 (m, 4H, 2"-H, 3"-H), 6.76 (td, ^3J = 7.4 Hz, ^3J = 6.6 Hz, ^4J = 2.3 Hz, 2H, 4"-H), 6.39 (d, ^3J = 8.0 Hz, 2H, 5"-H), 2.36 (s, 3H, CH_3), 2.18 (s, 6H, CH_3) ppm.

^{13}C NMR (101 MHz, CDCl$_3$) δ = 163.14 (C=O), 163.05 (C=O), 146.10 (C-1), 145.88 (2C, C-1"), 145.55 (C-5), 145.30 (C-4), 144.73 (C-3), 143.60 (C-6), 140.29 (C-4'), 135.94 (2C, C-2'), 134.21 (C-8), 133.86 (C-7), 132.65 (2C, C-6"), 129.67 (2C, C-3'), 129.47 (C-2), 126.33 (C-1'),

123.78 (2C, C-3"), 123.60 (2C, C-4"), 116.69 (2C, C-2"), 115.66 (2C, C-5"), 21.25 (CH_3), 18.18 (2C, CH_3) ppm.

MS (FAB, 3-NBA), m/z (%): 499 (32) [M+H]$^+$, 498 (48) [M]$^+$.

HRMS–FAB *(m/z)*: Calc. for [$C_{31}H_{22}O_3N_4$]$^+$: 498.1686; found: 498.1686.

IR (ATR, $\tilde{v}$) = 2921 (w), 2853 (w), 1790 (w), 1735 (vs), 1618 (w), 1571 (w), 1485 (vs), 1465 (s), 1368 (vs), 1326 (s), 1310 (s), 1290 (s), 1269 (vs), 1207 (m), 1181 (m), 1136 (s), 1122 (s), 1106 (s), 1034 (m), 864 (m), 832 (s), 741 (vs), 714 (m), 671 (m), 565 (m), 415 (m) cm^{-1}.

R$_f$ = 0.49 (Cyclohexane : ethyl acetate = 4:1).

6-(4-(10*H*-phenoxazin-10-yl)phenyl)-2-mesityl-1*H*-pyrrolo[3,4-*b*]quinoxaline-1,3(2*H*)-dione (68)

A microwave vial was charged with 6-bromo-2-mesityl-1*H*-pyrrolo[3,4-*b*]quinoxaline-1,3(2*H*)-dione (50.0 mg, 126 µmol, 1.00 equiv.), (4-phenoxazin-10-ylphenyl) boronic acid (45.9 mg, 151 µmol, 1.20 equiv.), K_2CO_3 (34.9 mg, 252 µmol, 2.00 equiv.) and Pd(PPh$_3$)$_4$ (19.4 mg, 16.8 µmol, 0.13 equiv.). Then, the reaction vial was sealed, evacuated, and backfilled with argon. Subsequently, toluene (2.00 mL) and water (500 µL) were added, and the reaction mixture was stirred at 100 °C, max. 150 W, without additional cooling for 15 min. After the reaction was done, the crude mixture was poured into an excess of water and extracted with dichloromethane (3×). The combined organic extracts were washed with brine, dried over Na_2SO_4 and concentrated under reduced pressure. The orange-yellow solid crude mixture was purified employing column chromatography (cyclohexane : dichloro-methane = 4:1 → 1:6) and flash chromatography using a cyclohexane/ethyl acetate gradient. The product (49.0 mg, 85.3 µmol, 68% yield) was obtained as a yellow-orange solid that is red in solution. The rotamers are indicated by the indices *a* and *b*.

¹H NMR (400 MHz, CDCl₃) δ = 8.73 (d, ^4J = 2.1 Hz, 1H, 2-H), 8.58 (d, ^3J = 8.9 Hz, 1H, 7-H), 8.38 (dd, ^3J = 8.9 Hz, ^4J = 2.1 Hz, 1H, 8-H), 8.04 (d, ^3J = 8.1 Hz, 2H, 2"-Hₐ), 7.58 (d, ^3J = 8.1 Hz, 2H, 3"-Hₐ), 7.17 (d, ^3J = 8.4 Hz, 1H, 2"-H_b), 7.07 (s, 2H, 3'-H), 7.03 (d, ^3J = 8.4 Hz, 1H, 3"-H_b), 6.74 (dd, ^3J = 7.5 Hz, ^4J = 1.8 Hz, 2H, 6"-Hₐ), 6.70 (dd, ^3J = 7.5 Hz, ^4J = 1.6 Hz, 2H, 7"-Hₐ), 6.68–6.61 (m, 4H, 6"-H_b , 7"-H_b , 8"-Hₐ), 6.61–6.55 (m, 1H, 8"-H_b), 6.05 (dd, ^3J = 7.8 Hz, ^4J = 1.6 Hz, 2H, 9"-Hₐ), 5.92 (dd, ^3J = 7.5 Hz, ^4J = 1.9 Hz, 1H, 9"-H_b), 2.37 (s, 3H, C*H*₃), 2.20 (s, 6H, C*H*₃) ppm.

¹³C NMR (101 MHz, CDCl₃) δ = 163.28 (2C, C=O), 155.88 (C_b-4"), 145.72 (C-4), 145.33 (C-1), 145.10 (C-3), 144.88 (C-5), 144.19 (2C, C_b-5"), 144.14 (2C, Cₐ-5"), 144.10 (C-6), 140.47 (Cₐ-4"), 140.34 (C-4'), 138.45 (Cₐ-1"), 136.01 (2C, C-2'), 134.88 (2C, C_b-10"), 134.16 (2C, Cₐ-10"), 133.16 (C-8), 132.14 (2C, Cₐ-3"), 132.11 (2C, C_b-2"), 132.00 (C-7), 131.50 (C_b-1"), 130.46 (2C, Cₐ-2"), 129.72 (C-3'), 128.63 (C-2), 126.40 (C-1'), 123.44 (2C, Cₐ-8"), 123.33 (C_b-8"), 121.85 (2C, Cₐ-7"), 121.22 (2C, C_b-7"), 117.92 (C_b-3"), 115.81 (2C, Cₐ-6"), 115.42 (2C, C_b-6"), 113.44 (2C, Cₐ-9"), 113.32 (2C, C_b-9"), 21.30 (2C, C*H*₃), 18.27 (C*H*₃) ppm.

HRMS–FAB *(m/z)*: Calc. for $[C_{37}H_{26}O_3N_4]^+$: 574.1999; found: 574.1998.

IR (ATR, ṽ) = 2953 (w), 2921 (w), 2850 (w), 1786 (w), 1724 (vs), 1626 (w), 1589 (w), 1497 (m), 1483 (vs), 1462 (s), 1400 (w), 1375 (s), 1329 (vs), 1292 (s), 1268 (vs), 1204 (m), 1137 (s), 1125 (vs), 1111 (s), 1038 (m), 870 (m), 837 (m), 826 (m), 815 (s), 734 (vs), 713 (s), 616 (m), 565 (s), 418 (m) cm⁻¹.

R_f = 0.68 (Cyclohexane : ethyl acetate = 2:1).

5.2.4. Red-Shifted Azo-Menthol Derivatives

(1R,2S,5R)-2-isopropyl-5-methyl-N-(4-((E)-phenyldiazenyl)phenyl)cyclohexane-1-carboxamide (69)

The synthesis of this compound had been first established by Dr. David B. Konrad.

A flask was charged with (1R,2S,5R)-2-isopropyl-5-methyl-cyclohexanecarboxylic acid (150 mg, 814 μmol, 1.00 equiv.) and evacuated and backfilled with nitrogen three times. Then, dichloromethane (2.00 mL, abs.) was added, and oxalyl chloride (124 mg, 83.8 μL, 977 μmol, 1.20 equiv.) was added dropwise while stirring at room temperature for 3 h. Subsequently, all volatile components were removed under reduced pressure. Then, DMAP (9.94 mg, 81.4 μmol, 0.10 equiv.) and 4-aminoazobenzene (177 mg, 895 μmol, 1.10 equiv.) were added, and the flask was evacuated and backfilled with nitrogen two times. Subsequently, toluene (2.00 mL, abs.) was added and the mixture cooled to 0 °C. After 5 min, DIPEA (158 mg, 1.22 mmol, 1.50 equiv.) was added dropwise, and the reaction mixture was stirred for 48 h. Then, the reaction mixture was poured into an excess of water and extracted with ethyl acetate. The combined organic phases were washed with 1 M HCl, water, and brine and dried over Na_2SO_4. After filtration and concentration under reduced pressure, the crude product was purified employing flash chromatography using a hexane/ethyl acetate gradient. The product (247 mg, 680 μmol, 83% yield) was obtained as an orange solid.

^{1}H NMR (400 MHz, CDCl$_3$) δ = 7.95–7.88 (m, 4H, 3'-H, 6'-H), 7.73–7.68 (m, 2H, 2'-H), 7.55–7.48 (m, 2H, 7'-H), 7.48–7.42 (m, 1H, 8'-H), 7.30 (s, 1H, NH), 2.19 (td, J = 11.3 Hz, J = 3.5 Hz, 1H, 1-H), 1.94 (dq, J = 12.5 Hz, J = 2.6 Hz, 1H, 6e-H), 1.84 (qd, J = 6.9 Hz, J = 2.6 Hz, 1H, 7-H), 1.80–1.71 (m, 2H, 3-H, 4-H), 1.66 (tt, J = 11.4 Hz, J = 3.0 Hz, 1H, 2-H), 1.49–1.38 (m, 1H, 5-H), 1.31 (dt, J = 23.1 Hz, J = 11.4 Hz, 1H, 6a-H), 1.15–0.98 (m, 2H, 3-H, 4-H), 0.94 (dd, J = 6.6 Hz, J = 5.0 Hz, 6H, CH_3, CH(CH_3)$_2$), 0.85 (d, J = 6.9 Hz, 3H, CH(CH_3)$_2$) ppm.

^{13}C NMR (101 MHz, CDCl$_3$) δ = 174.53 (C=O), 152.84 (C-5'), 149.10 (C-4'), 140.69 (C-1'), 130.88 (+, C-8'), 129.21 (+, 2C, C-7'), 124.17 (+, 2C, C-3'), 122.88 (+, 2C, C-6'), 119.86 (+, 2C, C-2'), 51.18 (+, C-1), 44.77 (+, C-2), 39.64 (–, C-6), 34.63 (–, C-4), 32.49 (+, C-5), 29.03 (+, C-7), 24.10 (–, C-3), 22.44 (+, CH_3), 21.55 (+, CH(CH$_3$)$_2$), 16.47 (+, CH(CH$_3$)$_2$) ppm.

MS (FAB, 3-NBA), m/z (%): 364 (100) [M+H]$^+$, 363 (31) [M]$^+$.

HRMS–FAB *(m/z)*: Calc. for [C$_{23}$H$_{30}$ON$_3$]$^+$: 364.2383; found: 364.2382.

IR (ATR, ṽ) = 3296 (w), 2952 (w), 2914 (m), 2870 (w), 2846 (w), 1656 (vs), 1594 (s), 1517 (vs), 1442 (m), 1405 (vs), 1387 (m), 1371 (m), 1340 (w), 1299 (m), 1251 (m), 1181 (m), 1145 (m), 1010 (w), 921 (w), 840 (vs), 765 (s), 715 (m), 686 (vs), 650 (m), 550 (vs), 527 (w) cm^{-1}.

R$_f$ = 0.55 (Cyclohexane : ethyl acetate = 4:1).

(1R,2S,5R)-2-Isopropyl-5-methyl-N-(4-((E)-(2,4,6-trimethoxyphenyl)diazenyl)phenyl)cyclohexane-1-carboxamide (70a)

A sealable vial was charged with (1R,2S,5R)-2-isopropyl-5-methyl-cyclohexanecarboxylic acid (5.77 mg, 31.3 μmol, 1.00 equiv.) and EDC·HCl (6.61 mg, 34.5 μmol, 1.10 equiv.). It was then sealed and evacuated, and backfilled with argon three times. Then, dichloromethane (300 μL, abs.) was added and the mixture was stirred at 0 °C for 30 min. A separate vial was charged with 4-((2,4,6-trimethoxyphenyl)diazenyl)aniline (9.00 mg, 31.3 μmol, 1.00 equiv.) and DMAP (3.83 mg, 31.3 μmol, 1.00 equiv.) and evacuated and backfilled with argon. Subsequently, dichloromethane (300 μL, abs.) was added. The mixture was stirred at room temperature for 30 min. Subsequently, the amine solution was added dropwise to the O-acylisourea-containing solution while stirring at 0 °C. The reaction mixture was stirred at room temperature for 6.5 d. After the reaction was stopped, the solvent was removed under reduced pressure to give the crude product. The crude product was purified employing flash chromatography using a hexane/ethyl acetate gradient. The (E/Z)-product mixture (3.00 mg, 6.61 μmol, 21% yield) was obtained as an orange-red solid. The NMR data stated below refers to the dominant isomer. However, it is noted that the signals belonging to the (−)-menthyl moiety were difficult to distinguish.

^{1}H NMR (400 MHz, CDCl$_3$) δ = 7.87 (d, ^3J = 8.7 Hz, 2H, 3'-H), 7.67 (d, ^3J = 8.6 Hz, 2H, 2'-H), 6.23 (s, 2H, 7'-H), 3.88 (s, 9H, OC*H$_3$*), 2.18 (td, J = 11.3 Hz, J = 3.4 Hz, 1H, 1-H), 1.97–

1.54 (m, 5H, 2-H, 5-H, 7-H, H$_{menthyl}$), 1.53–1.29 (m, 2H, 6-H, H$_{menthyl}$), 1.11–0.97 (m, 2H, H$_{menthyl}$), 0.93 (dd, J = 6.6 Hz, J = 3.3 Hz, 6H, CH_3), 0.84 (d, J = 6.9 Hz, 3H, CH_3) ppm.

^{13}C NMR (101 MHz, CDCl$_3$) δ = 174.45, 155.50, 153.34, 140.02, 124.04, 123.44, 119.81, 119.09, 91.80, 91.45, 56.76, 56.27, 55.76, 51.12, 44.75, 39.63, 34.66, 32.48, 29.85, 28.99, 24.12, 22.45, 22.43, 21.55, 21.52, 16.48 ppm.

MS (FAB, 3-NBA), m/z (%): 454 (9) [M+H]$^+$.

HRMS–FAB *(m/z)*: Calc. for [C$_{26}$H$_{36}$O$_4$N$_3$]$^+$: 454.2700; found: 454.2700.

R$_f$ = 0.26 (Cyclohexane : ethyl acetate = 2:1).

(1R,2S,5R)-2-isopropyl-5-methyl-N-(4-((E)-(3,4,5-trimethoxyphenyl)diazenyl)phenyl) cyclohexane-1-carboxamide (70b)

To a flame-dried flask, (1R,2S,5R)-2-isopropyl-5-methyl-cyclohexanecarboxylic acid (19.2 mg, 104 μmol, 1.50 equiv.) in dichloromethane (500 μL, abs.) was added. Subsequently, the solution was cooled to 0 °C, and oxalyl chloride (148 mg, 100 μL, 1.17 mmol, 16.80 equiv.) was added dropwise while stirring. The reaction mixture was stirred for 10 min at 0 °C and subsequently stirred at room temperature for 3 h. Then, all volatile components were removed under reduced pressure, and the flask was evacuated and backfilled with nitrogen. Dichloromethane (500 μL, abs.) was added, and the acyl chloride-containing solution was cooled to 0 °C. Meanwhile, in a separate vial, (E)-4-((3,4,5-trimethoxyphenyl)diazenyl)aniline (20.0 mg, 69.6 μmol, 1.00 equiv.) in dichloromethane (500 μL, abs.) under nitrogen atmosphere was cooled to 0 °C. Subsequently, pyridine (39.3 mg, 40.0 μL, 497 μmol, 7.13 equiv) was added and the mixture was stirred at 0 °C for 15 min. Subsequently, the amine-containing solution was dropwise added to the acyl chloride-containing solution at 0 °C and stirred at room temperature for 18 h. Then, the reaction mixture was poured into 1 M HCl and extracted with dichloromethane two times. The combined organic phases were washed with NaHCO$_3$ and brine, dried over Na$_2$SO$_4$, and concentrated under reduced pressure. The crude mixture was purified employing column chromatography (hexane : ethyl acetate = 50:1 → 20:1 → 10:1). The product (18.0 mg, 39.7 µmol, 57% yield) was obtained as a dark red oil.

^{1}H NMR (400 MHz, CDCl$_3$) δ = 7.93–7.85 (m, 2H, 3'-H), 7.72 (d, ^3J = 8.6 Hz, 2H, 2'-H), 7.51 (s, 1H, NH), 7.23 (s, 2H, 6'-H), 3.96 (s, 6H, OCH_3), 3.93 (s, 3H, OCH_3), 2.20 (td, J = 11.2 Hz, J = 3.5 Hz, 1H, 1-H), 1.95–1.86 (m, 1H, 6-H), 1.82 (td, J = 6.9 Hz, J = 2.5 Hz, 1H, 7-H), 1.80–1.68 (m, 2H, 3-H, 4-H), 1.65 (tt, J = 11.4 Hz, J = 2.9 Hz, 1H, 2-H), 1.44–1.37 (m, 1H, 5-H), 1.36–1.28 (m, 1H, 6-H), 1.09–0.97 (m, 2H, 3-H, 4-H), 0.92–0.85 (m, 6H, CH_3, CH(CH_3)$_2$), 0.84 (d, J = 6.9 Hz, 3H, CH(CH_3)$_2$) ppm.

^{13}C NMR (101 MHz, CDCl$_3$) δ = 174.72 (C=O), 153.64 (C-8'), 148.91 (C-5'), 148.69 (C-4'), 140.65 (2C, C-7'), 140.53 (C-1'), 123.99 (+, 2C, C-3'), 119.95 (+, 2C, C-2'), 100.40 (+, 2C, C-6'), 61.17 (+, OCH_3), 56.32 (+, 2C, OCH_3), 51.05 (+, C-1), 44.70 (+, C-2), 39.63 (−, C-6), 34.60 (−, C-4), 32.44 (+, C-5), 29.01 (+, C-7), 24.06 (−, C-3), 22.42 (+, CH_3), 21.54 (+, CH(CH_3)$_2$), 16.44 (+, CH(CH_3)$_2$) ppm.

MS (EI, 70 eV, 170 °C), m/z (%): 453 (100) [M]$^+$.

HRMS–EI (m/z): Calc. for [C$_{26}$H$_{35}$O$_4$N$_3$]$^+$: 453.2622; found: 453.2622.

IR (ATR, ṽ) = 3305 (w), 2951 (w), 2922 (m), 2870 (w), 1662 (m), 1595 (s), 1528 (s), 1497 (vs), 1465 (s), 1411 (vs), 1329 (s), 1300 (vs), 1235 (s), 1221 (s), 1174 (m), 1149 (s), 1126 (vs), 1001 (s), 909 (w), 847 (vs), 783 (w), 728 (s), 691 (m), 639 (m), 526 (m), 503 (w) cm^{-1}.

R$_f$ = 0.29 (Cyclohexane : ethyl acetate = 4:1).

(1R,2S,5R)-N-(4-((E)-(4-(diethylamino)phenyl)diazenyl)phenyl)-2-isopropyl-5-methylcyclohexane-1-carboxamide (70c)

A flask was charged with (1R,2S,5R)-2-isopropyl-5-methyl-cyclohexanecarboxylic acid (8.00 mg, 43.4 µmol, 1.06 equiv.) and evacuated and backfilled with nitrogen three times. Then, dichloromethane (1.00 mL, abs.) was added and oxalyl dichloride (74.0 mg,

50.0 µL, 583 µmol, 14.20 equiv.) was added dropwise while stirring at room temperature for 3 h. Subsequently, all volatile components were removed under reduced pressure. Then, 4-((4-aminophenyl)diazenyl)-*N*,*N*-diethylaniline (11.0 mg, 41.0 µmol, 1.00 equiv.) was added and the flask was evacuated and backfilled with nitrogen two times. After adding dichloromethane (200 µL, abs.), pyridine (13.0 mg, 13.2 µL, 164 µmol, 4.00 equiv.) was added dropwise and the reaction mixture was stirred for 24 h. Then, the reaction mixture was poured into an excess of water and extracted with ethyl acetate. The combined organic extracts were washed with 1 M HCl, water and brine, and dried over Na_2SO_4. After filtration and concentration under reduced pressure, the crude product was purified employing column chromatography (hexane : ethyl acetate = 100:1 → 20:1). The product (7.00 mg, 16.1 µmol, 39% yield) was obtained as a highly viscous orange oil.

^{1}H NMR (400 MHz, $CDCl_3$) δ = 7.83 (dq, J = 7.9 Hz, J = 3.1 Hz, 4H, 6'-H, 3'-H), 7.68–7.61 (m, 2H, 2'-H), 7.19 (s, 1H, N*H*), 6.75–6.68 (m, 2H, 7'-H), 3.45 (q, J = 7.1 Hz, 4H, N(C*H*$_2$CH$_3$)$_2$), 2.16 (td, J = 11.2 Hz, J = 3.4 Hz, 1H, 1-H), 1.97–1.89 (m, 1H, 6-H), 1.88–1.81 (m, 1H, 7-H), 1.80–1.71 (m, 2H, 3-H, 4-H), 1.69–1.60 (m, 1H, 2-H), 1.45–1.36 (m, 1H, 5-H), 1.35–1.27 (m, 1H, 6-H), 1.23 (t, J = 7.1 Hz, 6H, N(CH$_2$C*H*$_3$)$_2$), 1.14 – 0.97 (m, 2H, 3-H, 4-H), 0.93 (dd, J = 6.6 Hz, J = 3.8 Hz, 6H, C*H*$_3$), 0.84 (d, J = 6.9 Hz, 3H, C*H*$_3$).

^{13}C NMR (101 MHz, $CDCl_3$) δ = 174.29 (C=O), 150.11 (C-8'), 149.82 (C-4'), 143.27 (C-5'), 139.02 (C-1'), 125.27 (+, 2C, C-3'), 123.21 (+, 2C, C-6'), 119.91 (+, 2C, C-2'), 111.15 (+, 2C, C-7'), 51.17 (+, C-1), 44.85 (–, 2C, N(CH$_2$CH$_3$)$_2$), 44.76 (+, C-2), 39.65 (–, C-6), 34.67 (–, C-4), 32.50 (+, C-5), 28.97 (+, C-7), 24.11 (–, C-3), 22.46 (+, *C*H$_3$), 21.56 (+, CH(*C*H$_3$)$_2$), 16.46 (+, CH(*C*H$_3$)$_2$), 12.83 (+, 2C, N(CH$_2$*C*H$_3$)$_2$).

MS (FAB, 3-NBA), m/z (%): 435 (100) [M+H]$^+$, 434 (96) [M]$^+$.

HRMS–FAB *(m/z)*: Calc. for [$C_{27}H_{38}ON_4$]$^+$: 434.3040; found: 434.3038.

IR (ATR, ṽ) = 3276 (w), 3242 (w), 2952 (m), 2922 (m), 2868 (m), 1664 (w), 1647 (w), 1594 (vs), 1513 (vs), 1445 (m), 1421 (m), 1405 (s), 1392 (vs), 1375 (s), 1354 (vs), 1316 (s), 1268 (vs), 1245 (vs), 1196 (m), 1149 (vs), 1137 (vs), 1096 (m), 1077 (s), 1010 (m), 844 (s), 815 (vs), 790 (m), 731 (m), 704 (m), 550 (s), 523 (m), 504 (m), 395 (m) cm^{-1}.

R$_f$ = 0.48 (Cyclohexane : ethyl acetate = 4:1).

(E)-4-((2,4,6-trimethoxyphenyl)diazenyl)aniline (72a)

H₂N, 1
   OMe
2   4 N=N
3   5
MeO 6 7 8 OMe

Na₂S (54.1 mg, 693 μmol, 2.20 equiv.) was added to a solution of (E)-1-(4-nitrophenyl)-2-(2,4,6-trimethoxyphenyl) diazene (100 mg, 315 μmol, 1.00 equiv.) in 1,4-dioxane (1.00 mL) and water (1.00 mL). The mixture was refluxed at 80 °C for 24 h. After cooling to room temperature, all volatile components were removed under reduced pressure. The mixture was diluted in ethyl acetate and washed with 1 M NaOH, sat. NaHCO₃ solution and brine before being dried over Na₂SO₄. The crude mixture was purified employing column chromatography (hexane : ethyl acetate = 10:1→ 2:1 + 1% NEt₃). The product (36.0 mg, 125 μmol, 40% yield) was obtained as an orange-red solid.

^{1}H NMR (400 MHz, CDCl₃) δ = 7.81–7.70 (m, 2H, 3-H), 6.79–6.67 (m, 2H, 2-H), 6.23 (s, 2H, 7-H), 3.86 (s, 3H, CH_3), 3.84 (s, 6H, CH_3) ppm.

^{13}C NMR (101 MHz, CDCl₃) δ = 160.94 (C-8), 154.35 (2C, C-6), 149.04 (C-1), 146.84 (C-4), 128.19 (C-5), 124.59 (2C, C-3), 114.74 (2C, C-2), 91.88 (2C, C-7), 56.66 (2C, CH_3), 55.62 (CH_3) ppm.

(E)-1-(4-nitrophenyl)-2-(2,4,6-trimethoxyphenyl)diazene (74a)

O₂N, 1
   OMe
2   4 N=N
3   5
MeO 6 7 8 OMe

A sealable vial was charged with 4-nitrobenzenediazonium tetrafluoroborate (338 mg, 1.43 mmol, 1.20 equiv.), 1,3,5-trimethoxybenzene (200 mg, 1.19 mmol, 1.00 equiv.) and Na₂SO₄ (169 mg, 1.19 mmol, 1.00 equiv.). Subsequently, it was evacuated and backfilled with nitrogen three times. Then, methanol (15.00 mL, abs.) was added, and the reaction mixture was stirred at 0 °C for 3 d. After the reaction was completed, the reaction mixture was diluted with dichloromethane and washed with NaHCO₃ solution. The organic phase was then washed with brine, dried over Na₂SO₄, and concentrated under reduced pressure. The product (343 mg, 1.08 mmol, 91% yield) was obtained as a deep red solid.

^{1}H NMR (400 MHz, DMSO-d₆) δ = 8.37 (d, ^3J = 8.6 Hz, 2H, 2-H), 7.84 (d, ^3J = 8.6 Hz, 2H, 3-H), 6.40 (s, 2H, 7-H), 3.91 (s, 3H, CH₃), 3.85 (s, 6H, CH₃) ppm.

^{13}C NMR (101 MHz, DMSO-d$_6$) δ = 164.04 (C-8), 156.97 (C-4), 155.85 (2C, C-6), 147.28 (C-1), 126.76 (C-5), 124.94 (2C, C-2), 122.41 (2C, C-3), 91.59 (2C, C-7), 56.36 (2C, CH$_3$), 55.88 (CH$_3$) ppm.

MS (FAB, 3-NBA), m/z (%): 318 (100).

HRMS–FAB *(m/z)*: Calc. for [C$_{15}$H$_{16}$O$_5$N$_3$]$^+$: 318.1084; found: 318.1086.

IR (ATR, ṽ) = 1595 (vs), 1584 (vs), 1570 (s), 1514 (vs), 1415 (s), 1309 (vs), 1286 (vs), 1232 (vs), 1211 (vs), 1160 (vs), 1142 (vs), 1118 (vs), 1101 (vs), 1060 (vs), 1020 (vs), 928 (s), 856 (vs), 850 (vs), 819 (vs), 812 (vs), 758 (vs), 738 (s), 686 (vs), 554 (s), 516 (vs), 439 (vs), 392 (vs) cm^{-1}.

EA: Calc. for C$_{15}$H$_{15}$N$_3$O$_5$: C 56.78; H 4.76; N 13.24. Found: C 56.46; H 4.77; N 13.00%.

R$_f$ = 0.12 (Cyclohexane : ethyl acetate = 4:1).

(*E*)-1-(4-nitrophenyl)-2-(3,4,5-trimethoxyphenyl)diazene

A sealable vial was charged with 4-nitrobenzenediazonium tetrafluoroborate (338 mg, 1.43 mmol, 1.10 equiv.) and 2,6-dimethoxyphenol (200 mg, 1.30 mmol, 1.00 equiv.). Subsequently, it was evacuated and backfilled with nitrogen three times. Then, methanol (15.00 mL) was added, and the reaction mixture was stirred at 0 °C for 22 h. After the reaction was completed, the reaction mixture was diluted with dichloromethane and washed with NaHCO$_3$ solution. The organic phase was then washed with brine, dried over Na$_2$SO$_4$, and concentrated under reduced pressure. The formation of 2,6-dimethoxy-4-((4-nitrophenyl)diazenyl)phenol was confirmed by LCMS analysis, and the crude product was further reacted without additional purification.

To a flask with 2,6-dimethoxy-4-((4-nitrophenyl)diazenyl)phenol (117 mg, 386 μmol, 1.00 equiv.) in acetone (3.50 mL), K$_2$CO$_3$ (80.0 mg, 579 μmol, 1.50 equiv.) was added while stirring at 0 °C. After 15 min, iodomethane (548 mg, 240 μL, 3.86 mmol, 10.00 equiv.) was added dropwise, and the reaction was stirred at 40 °C for 24 h. Subsequently, all volatile components were removed, and the crude mixture was purified employing flash chromatography using a hexane/ethyl acetate gradient. The product (93.0 mg, 293 μmol, 76% yield) was obtained as a bright orange solid.

¹H NMR (400 MHz, CDCl₃) δ = 8.38 (d, ^3J = 8.9 Hz, 2H, 2-H), 8.09–7.92 (m, 2H, 3-H), 7.30 (s, 2H, 6-H), 3.98 (s, 6H, C*H*₃), 3.97 (s, 3H, C*H*₃) ppm.

¹³C NMR (101 MHz, CDCl₃) δ = 155.87 (C-4), 153.76 (2C, C-7), 148.68 (C-1), 148.38 (C-5), 142.17 (C-8), 124.91 (2C, C-2), 123.43 (2C, C-3), 101.30 (2C, C-6), 61.28 (*C*H₃), 56.43 (2C, *C*H₃) ppm.

MS (FAB, 3-NBA), m/z (%): 318 (64) [M+H]⁺, 317 (53) [M]⁺.

HRMS–FAB *(m/z)*: Calc. for [C₁₅H₁₅O₅N₃]⁺: 317.1006; found: 317.1007.

IR (ATR, ṽ) = 3102 (w), 2941 (m), 2918 (m), 2833 (m), 1594 (s), 1519 (s), 1494 (vs), 1463 (s), 1414 (s), 1334 (vs), 1306 (vs), 1222 (vs), 1122 (vs), 1103 (vs), 1096 (vs), 992 (vs), 864 (vs), 840 (vs), 823 (vs), 755 (s), 693 (vs) cm⁻¹.

R_f_ = 0.12 (Cyclohexane : ethyl acetate = 4:1).

_N_-((1_R_,2_S_,5_R_)-2-isopropyl-5-methylcyclohexyl)-4-((_E_)-(3,4,5-trimethoxyphenyl)diazenyl) benzamide (77)

A sealable vial was charged with 4-((3,4,5-trimethoxyphenyl)diazenyl)benzoic acid (11.2 mg, 35.4 μmol, 1.10 equiv) and HATU (14.7 mg, 38.6 μmol, 1.20 equiv). It was then sealed, evacuated, and backfilled with nitrogen. Then, *N,N*-dimethylformamide (403 μL, abs.) was added and the mixture was stirred at 0 °C for 30 min. In a separate vial, a mixture of (2*S*,5*R*)-5-methyl-2-propan-2-ylcyclohexan-1-amine (5.00 mg, 32.2 μmol, 1.00 equiv) and DIPEA (6.24 mg, 8.41 μL, 48.3 μmol, 1.50 equiv) was stirred in *N,N*-dimethylformamide (200 μL) at 0 °C. Subsequently, the amine-containing mixture was added to the carboxylic acid-containing mixture, and the resulting reaction mixture was stirred for 24 h at room temperature. Then, the mixture was poured into an excess of water and extracted with dichloromethane three times. The combined organic phases were extracted with brine, dried over Na_2SO_4, and concentrated under reduced pressure. The crude mixture was purified employing column chromatography (hexane : ethyl acetate = 10:1 → 5:1). The product (7.00 mg, 15.4 μmol, 48% yield) was obtained as an orange solid.

^{1}**H NMR** (400 MHz, CDCl$_3$) δ = 7.92 (q, ^3J = 8.7 Hz, 4H, 2'-H, 3'-H), 7.28 (s, 2H, 6'-H), 5.84 (d, J = 9.3 Hz, 1H, NH), 4.09–3.99 (m, 1H, 1-H), 3.98 (s, 6H, OCH_3), 3.95 (s, 3H, OCH_3), 2.16–2.07 (m, 1H, 6-H), 1.99 (dtd, J = 14.2 Hz, J = 7.1 Hz, J = 2.0 Hz, 1H, 7-H), 1.75 (td, J = 12.0 Hz, J = 10.7 Hz, J = 3.3 Hz, 2H, 3-H, 4-H), 1.61–1.51 (m, 1H, 5-H), 1.23–1.17 (m, 2H, 2-H, 3-H), 0.96–0.89 (m, 8H, 4-H, 6-H, CH_3, CH(CH_3)$_2$), 0.89–0.85 (m, 3H, CH(CH_3)$_2$) ppm.

^{13}**C NMR** (101 MHz, CDCl$_3$) δ = 166.10 (C=O), 154.21 (C-4'), 153.71 (2C, C-7'), 148.54 (C-5'), 141.34 (C-8'), 136.82 (C-1'), 127.95 (+, 2C, C-2'), 122.95 (+, 2C, C-3'), 100.87 (+, 2C, C-6'), 61.23 (+, OCH_3), 56.39 (+, 2C, OCH_3), 50.75 (+, C-1), 48.59 (+, C-2), 43.35 (−, C-6), 34.70 (−, C-4), 32.08 (+, C-5), 27.25 (+, C-7), 24.07 (−, C-3), 22.31 (+, CH_3), 21.37 (+, CH(CH_3)$_2$), 16.43 (+, CH(CH_3)$_2$) ppm.

MS (EI, 70 eV, 170 °C), m/z (%): 453 (19) [M]$^+$.

HRMS–EI *(m/z)*: Calc. for [C$_{26}$H$_{35}$O$_4$N$_3$]$^+$: 453.2622; found: 453.2620.

IR (ATR, ṽ) = 3309 (vw), 2921 (m), 2851 (w), 1630 (s), 1596 (m), 1537 (s), 1494 (vs), 1466 (s), 1455 (s), 1429 (m), 1412 (s), 1330 (s), 1302 (vs), 1273 (m), 1234 (s), 1220 (vs), 1179 (w), 1125 (vs), 1037 (w), 1003 (s), 860 (s), 844 (s), 771 (w), 722 (m), 697 (w), 674 (w), 639 (m), 611 (m), 577 (w), 523 (w) cm^{-1}.

R$_f$ = 0.26 (Cyclohexane : ethyl acetate = 4:1).

(2S,5R,E)-2-Isopropyl-5-methylcyclohexan-1-one oxime (78)

The synthesis was carried out according to a literature procedure.[76] L-menthone (925 mg, 1.03 mL, 6.00 mmol, 1.00 equiv.) was dissolved in ethanol (7.50 mL) and water (7.50 mL). While stirring, hydroxylamine hydrochloride (1.25 g, 18.0 mmol, 3.00 equiv.) and NaOH (1.44 g, 36.0 mmol, 6.00 equiv.) were added. The reaction mixture was stirred at room temperature for 63 h. Subsequently, all volatile components were removed under reduced pressure and the crude mixture was diluted with water. The aqueous solution was then extracted with ethyl acetate (3×). The combined organic extracts were washed with brine, dried over Na_2SO_4 and concentrated under reduced pressure. The crude mixture was further purified employing column chromatography (cyclohexane : ethyl acetate = 5:1). The product (970 mg, 5.73 mmol, 96% yield) was obtained as a colorless, crystalline solid.

^{1}H NMR (400 MHz, CDCl$_3$) δ 3.05–2.95 (m, 1H), 2.12 (dq, J = 13.3, 6.7 Hz, 1H), 1.93–1.69 (m, 5H), 1.38 (dtd, J = 13.1, 10.3, 2.9 Hz, 1H), 1.23–1.09 (m, 1H), 0.98 (d, J = 6.2 Hz, 3H), 0.92 (t, J = 6.4 Hz, 6H) ppm.

(E)-4-((2,4,6-trimethoxyphenyl)diazenyl)benzoic acid (79a)

A flask was charged with methyl (E)-4-((2,4,6-trimethoxy-phenyl)diazenyl)benzoate (180 mg, 545 μmol, 1.00 equiv.). Subsequently, methanol (5.00 mL) and 1 M NaOH (200 mg, 5.00 mL, 5.00 mmol, 9.18 equiv.) were added, and the reaction mixture was stirred at 65 °C for 3 h. Subsequently, 5 M HCl was added to adjust the pH to pH 2. The precipitate was filtered off and washed with diethyl ether. The product (172 mg, 544 μmol, quant. yield) was obtained as a red solid.

^{1}H NMR (400 MHz, DMSO-d$_6$) δ = 8.11–8.05 (m, 2H, 2-H), 7.76–7.67 (m, 2H, 3-H), 6.38 (s, 2H, 7-H), 3.88 (s, 3H, CH$_3$), 3.82 (s, 6H, CH$_3$) ppm.

^{13}C NMR (101 MHz, DMSO-d$_6$) δ = 166.90 (C=O), 163.02 (C-8), 155.75 (C-4), 155.12 (2C, C-6), 131.36 (C-1), 130.46 (2C, C-2), 126.78 (C-5), 121.62 (2C, C-3), 91.64 (2C, C-7), 56.33 (2C, CH$_3$), 55.80 (CH$_3$) ppm.

MS (FAB, 3-NBA), m/z (%): 317 (100) [M+H]$^+$.

HRMS–FAB *(m/z)*: Calc. for $[C_{16}H_{17}O_5N_2]^+$: 317.1132; found 317.1131.

IR (ATR, ṽ) = 3446 (w), 3367 (w), 2837 (w), 1707 (m), 1697 (m), 1618 (s), 1605 (m), 1591 (m), 1541 (m), 1504 (m), 1485 (vs), 1449 (s), 1417 (s), 1388 (m), 1371 (m), 1332 (vs), 1306 (s), 1261 (m), 1220 (vs), 1170 (vs), 1162 (vs), 1135 (vs), 1112 (vs), 1007 (m), 942 (m), 924 (s), 866 (m), 827 (vs), 815 (s), 772 (s), 759 (s), 688 (m), 650 (m), 606 (m), 552 (s), 518 (s), 504 (s), 465 (m), 419 (m) cm^{-1}.

R$_f$ = 0.02 (Cyclohexane : ethyl acetate = 2:1).

(*E*)-4-((3,4,5-trimethoxyphenyl)diazenyl)benzoic acid (79b)

A flask was charged with methyl (*E*)-4-((3,4,5-trimethoxyphenyl)diazenyl)benzoate (80.0 mg, 242 μmol, 1.00 equiv.). Subsequently, methanol (3.00 mL) and 1 M NaOH (120 mg, 3.00 mL, 3.00 mmol, 12.4 equiv.) were added, and the reaction mixture was stirred at 65 °C for 19 h. Subsequently, 5 M HCl was added to adjust the pH to pH 2. The mixture was then extracted with an ethyl acetate/methanol (9:1) mixture (3×). The combined organic extracts were dried over Na$_2$SO$_4$ and concentrated under reduced pressure. The product (76.0 mg, 240 μmol, 99% yield) was obtained as an orange solid.

^{1}H NMR (400 MHz, DMSO-d$_6$) δ = 8.17–8.09 (m, 2H, 2-H), 7.98–7.91 (m, 2H, 3-H), 7.31 (s, 2H, 6-H), 3.90 (s, 6H, C*H*$_3$), 3.78 (s, 3H, C*H*$_3$) ppm.

^{13}C NMR (101 MHz, DMSO-d$_6$) δ = 166.69 (C=O), 154.21 (C-4), 153.38 (2C, C-7), 147.74 (C-5), 140.98 (C-8), 132.57 (C-1), 130.63 (2C, C-2), 122.44 (2C, C-3), 100.67 (2C, C-6), 60.30 (C*H*$_3$), 56.07 (2C, C*H*$_3$) ppm.

MS (FAB, 3-NBA), m/z (%): 317 (46) [M+H]$^+$.

HRMS–FAB *(m/z)*: Calc. for $[C_{16}H_{17}O_5N_2]^+$: 317.1132; found: 317.1134.

IR (ATR, ṽ) = 3075 (w), 2946 (w), 2839 (w), 2649 (w), 2531 (w), 1680 (vs), 1594 (s), 1468 (s), 1412 (vs), 1302 (s), 1283 (vs), 1220 (vs), 1126 (vs), 1116 (vs), 990 (vs), 864 (vs), 846 (s), 775 (s), 715 (s), 696 (s), 552 (m), 524 (m) cm^{-1}.

R$_f$ = 0.04 (Cyclohexane : ethyl acetate = 4:1).

Methyl (*E*)-4-((2,4,6-trimethoxyphenyl)diazenyl)benzoate (82a)

A sealable vial was charged with 4-methoxycarbonyl benzene diazonium tetrafluoroborate (490 mg, 1.96 mmol, 1.20 equiv.) and 1,3,5-trimethoxybenzene (275 mg, 1.64 mmol, 1.00 equiv.) and Na$_2$SO$_4$ (232 mg, 1.64 mmol, 1.00 equiv.). Then, methanol (8.00 mL, abs.) and acetonitrile (3.40 mL, abs.) were added, and the mixture was stirred at 0 °C for 18 h. Subsequently, all volatile components were removed under reduced pressure. The crude mixture was purified employing flash chromatography using a hexane/ethyl acetate gradient. The product (287 mg, 869 μmol, 53% yield) was obtained as a red solid.

^{1}H NMR (400 MHz, CDCl$_3$) δ = 8.21–8.10 (m, 2H, 2-H), 7.87–7.80 (m, 2H, 3-H), 6.23 (s, 2H, 7-H), 3.94 (s, 3H, COOCH_3), 3.90 (s, 6H, OCH_3), 3.89 (s, 3H, OCH_3) ppm.

^{13}C NMR (101 MHz, CDCl$_3$) δ = 166.96 (C=O), 163.21 (C-8), 157.03 (C-4), 155.83 (2C, C-6), 130.69 (C-1), 130.64 (2C, C-2), 127.77 (C-5), 122.14 (2C, C-3), 91.54 (2C, C-7), 56.63 (2C, OCH_3), 55.70 (OCH_3), 52.33 (COOCH_3) ppm.

MS (FAB, 3-NBA), m/z (%): 331 (100) [M]$^+$.

HRMS–FAB *(m/z)*: Calc. for [C$_{17}$H$_{19}$O$_5$N$_2$]$^+$: 331.1288; found: 331.1290.

IR (ATR, ṽ) = 3002 (w), 2958 (w), 2921 (w), 2836 (w), 1706 (vs), 1595 (s), 1577 (vs), 1435 (vs), 1409 (s), 1404 (s), 1281 (vs), 1237 (s), 1224 (s), 1207 (vs), 1193 (vs), 1157 (vs), 1147 (vs), 1122 (vs), 1112 (vs), 1096 (vs), 1061 (vs), 1027 (vs), 1014 (s), 952 (s), 853 (vs), 809 (vs), 769 (vs), 698 (vs) cm^{-1}.

R$_f$ = 0.29 (Cyclohexane : ethyl acetate = 2:1).

Methyl (*E*)-4-((4-hydroxy-3,5-dimethoxyphenyl)diazenyl)benzoate (82b)

To a flask containing methyl 4-aminobenzoate (235 mg, 1.56 mmol, 1.20 equiv.) in methanol (5.00 mL), 5 M HCl (1.56 mL) was added, and the reaction mixture was cooled to 0°C. Subsequently, NaNO$_2$ (116 mg, 1.69 mmol, 1.30 equiv.) was dissolved in water (1.5 mL) and added dropwise to the beforementioned mixture. The mixture was allowed to stir for 30 min. A solution of 2,6-dimethoxyphenol (200 mg, 183 μL, 1.30 mmol, 1.00 equiv.) in methanol (5.00 mL) and 0.5 M K$_2$HPO$_4$ (5.00 mL) was prepared at 0 °C. Then, the diazonium solution was added dropwise onto the phenol mixture. The pH was maintained between 9-10 by adding 1 M KOH. Upon completion of the addition, the reaction mixture was allowed to stir for 23 h at 0 °C. Subsequently, the reaction was quenched by adjusting the pH to pH 4-6 with 2 M HCl and extracted with ethyl acetate (3×). The combined organic extracts were washed with brine, dried over Na$_2$SO$_4$, and concentrated under reduced pressure. The crude mixture was purified by flash chromatography using a hexane/ethyl acetate gradient. The product (278 mg, 879 μmol, 68% yield) was obtained as a red solid.

^{1}H NMR (400 MHz, CDCl$_3$) δ = 8.21–8.13 (m, 2H, 2-H), 7.94–7.86 (m, 2H, 3-H), 7.32 (s, 2H, 6-H), 6.00 (s, 1H, O*H*), 4.00 (s, 6H, C*H*$_3$), 3.95 (s, 3H, C*H*$_3$) ppm.

^{13}C NMR (101 MHz, CDCl$_3$) δ = 166.72 (C=O), 155.31 (C-4), 147.41 (2C, C-7), 145.72 (C-8), 138.72 (C-5), 131.37 (C-1), 130.74 (2C, C-2), 122.48 (2C, C-3), 101.01 (2C, C-6), 56.55 (2C, *C*H$_3$), 52.42 (*C*H$_3$) ppm.

MS (FAB, 3-NBA), m/z (%): 317 (100) [M+H]$^+$, 316 (57) [M]$^+$.

HRMS–FAB *(m/z)*: Calc. for [C$_{16}$H$_{17}$O$_5$N$_2$]$^+$: 317.1132; found: 317.1130.

IR (ATR, ṽ) = 3422 (w), 3006 (w), 2944 (w), 2842 (w), 1720 (s), 1604 (m), 1497 (m), 1466 (m), 1419 (s), 1310 (s), 1272 (vs), 1247 (s), 1197 (s), 1179 (s), 1095 (vs), 1033 (s), 870 (s), 857 (vs), 771 (vs), 694 (vs), 633 (s) cm^{-1}.

Methyl (*E*)-4-((3,4,5-trimethoxyphenyl)diazenyl)benzoate (82c)

To a solution of methyl (*E*)-4-((4-hydroxy-3,5-dimethoxyphenyl)diazenyl)benzoate (150 mg, 474 µmol, 1.00 equiv.) in acetone (3.00 mL), K_2CO_3 (98.3 mg, 711 µmol, 1.50 equiv.) was added, and the mixture was cooled to 0 °C. Subsequently, 18-crown-6 (18.8 mg, 15.2 µL, 71.1 µmol, 0.150 equiv.) and iodomethane (101 mg, 44.3 µL, 711 µmol, 1.50 equiv.) were added and the reaction was stirred for 25 h. After the reaction was done, all volatile components were removed under reduced pressure and the crude mixture was purified employing flash chromatography using a hexane/ethyl acetate gradient. The product (126 mg, 381 µmol, 80% yield) was obtained as an orange crystalline solid.

^{1}H NMR (400 MHz, $CDCl_3$) δ 8.23–8.15 (m, 2H, 2-H), 7.97–7.89 (m, 2H, 3-H), 7.29 (s, 2H, 6-H), 3.98 (s, 6H, C*H*₃), 3.96 (s, 3H, COOC*H*₃), 3.95 (s, 3H, C*H*₃) ppm.

^{13}C NMR (101 MHz, $CDCl_3$) δ = 166.69 (C=O), 155.24 (C-4), 153.72 (2C, C-7), 148.55 (C-5), 141.50 (C-8), 131.77 (C-1), 130.79 (2C, C-2), 122.66 (2C, C-3), 100.95 (2C, C-6), 61.24 (O*C*H₃), 56.40 (2C, O*C*H₃), 52.49 (COO*C*H₃) ppm.

MS (FAB, 3-NBA), m/z (%): 331 (100) [M+H]⁺, 330 (65) [M]⁺.

HRMS–FAB *(m/z)*. Calc. for $[C_{17}H_{19}O_5N_2]^+$: 331.1288; found: 331.1289.

IR (ATR, ṽ) = 2993 (w), 2938 (w), 2833 (w), 1718 (vs), 1594 (m), 1489 (m), 1412 (vs), 1273 (vs), 1222 (vs), 1196 (s), 1115 (vs), 1095 (vs), 996 (vs), 966 (s), 860 (vs), 841 (s), 833 (s), 771 (vs), 694 (vs), 639 (s) cm⁻¹.

R_f = 0.40 (Cyclohexane : ethyl acetate = 4:1).

Methyl (*E*)-4-((3,4-dihydroxyphenyl)diazenyl)benzoate (82d)

To a solution of methyl 4-aminobenzoate (329 mg, 2.18 mmol, 1.20 equiv.) in methanol (6.00 mL), 5 M HCl (199 mg, 1.09 mL, 5.45 mmol, 3.00 equiv.) was added, and the mixture was cooled to 0 °C. Then, 2 M NaNO$_2$ (150 mg, 2.18 mmol, 1.20 equiv.) was added dropwise, and the solution was stirred for 30 min. Separately, a solution of catechol (200 mg, 1.82 mmol, 1.00 equiv.) in methanol (8.00 mL) and 0.5 M K$_2$HPO$_4$ (8.00 mL) was prepared at 0°C. The diazonium solution was then added dropwise to the phenol mixture. The pH was maintained between 9-10 by adding 1 M KOH. Upon completion of the addition, the reaction mixture was allowed to stir for 13 h at 0 °C. The reaction was quenched by adjusting the pH to pH 4-6 with 5 M HCl and extracted with ethyl acetate (3×). The combined organic extracts were dried over Na$_2$SO$_4$ and concentrated under reduced pressure. The crude product was purified by flash chromatography using a dichloro-methane/methanol gradient. The product (100 mg, 367 µmol, 20% yield) was obtained as an orange-brown solid.

^{1}H NMR (400 MHz, DMSO-d$_6$) δ = 10.02 (s, 1H, O*H*), 9.58 (s, 1H, O*H*), 8.15–8.09 (m, 2H, 2-H), 7.92–7.85 (m, 2H, 3-H), 7.42 (dd, ^3J = 8.3 Hz, ^4J = 2.4 Hz, 1H, 10-H), 7.36 (d, ^4J = 2.4 Hz, 1H, 6-H), 6.95 (d, ^3J = 8.3 Hz, 1H, 9-H), 3.89 (s, 3H, C*H*$_3$) ppm.

^{13}C NMR (101 MHz, DMSO-d$_6$) δ = 165.73 (C=O), 154.80 (C-4), 150.74 (C-8), 146.21 (C-7), 145.63 (C-5), 130.45 (2C, C-2), 130.43 (C-1), 122.19 (2C, C-3), 120.62 (C-10), 115.44 (C-9), 106.02 (C-6), 52.33 (C*H*$_3$) ppm.

Methyl (*E*)-4-(benzo[*d*][1,3]dioxol-5-yldiazenyl)benzoate (82e)

A sealable vial was charged with methyl (*E*)-4-((3,4-dihydroxyphenyl)diazenyl)benzoate (50.0 mg, 184 µmol, 1.00 equiv.) and cesium carbonate (71.8 mg, 220 µmol, 1.20 equiv.). It was then evacuated and backfilled with nitrogen two times. Subsequently, *N,N*-dimethylformamide (1.00 mL, abs.) was added, and diiodo-methane (59.0 mg, 17.8 µL, 220 µmol, 1.20 equiv.) was added dropwise. The reaction mixture was stirred at 110 °C for 4 h. After cooling to room temperature, all volatile components were removed under reduced pressure. The mixture was extracted with water and ethyl acetate (3×). The combined organic extracts were washed with brine, dried over Na_2SO_4 and concentrated under reduced pressure. The crude mixture was further purified employing flash chromatography using a dichloromethane/methanol gradient. The product (6.00 mg, 21.1 μmol, 11% yield) was obtained as a dark red solid.

^{1}H NMR (400 MHz, CDCl$_3$) δ = 8.16–8.06 (m, 2H, 2-H), 7.88–7.79 (m, 2H, 3-H), 7.57 (dd, ^3J = 8.2 Hz, ^4J = 1.9 Hz, 1H, 10-H), 7.38 (d, ^4J = 1.9 Hz, 1H, 6-H), 6.90 (d, ^3J = 8.2 Hz, 1H, 9-H), 6.01 (s, 2H, CH_2), 3.88 (s, 3H, CH_3) ppm.

^{13}C NMR (101 MHz, CDCl$_3$) δ = 166.76 (C=O), 155.28 (C-4), 151.12 (C-8), 149.08 (C-9), 148.74 (C-5), 131.47 (C-1), 130.75 (2C, C-2), 125.02 (C-7), 122.59 (2C, C-3), 108.21 (C-6), 102.21 (C-10), 98.96 (*C*H$_2$), 52.45 (*C*H$_3$) ppm.

R$_f$ = 0.50 (Cyclohexane : ethyl acetate = 4:1).

(1R,2S,5R)-2-isopropyl-5-methylcyclohexyl 4-((E/Z)-(2,4,6-trimethoxyphenyl)diazenyl) benzoate (84a)

A flask was charged with 4-((2,4,6-trimethoxyphenyl)diazenyl)benzoic acid (20.2 mg, 64.0 μmol, 1.00 equiv) and DCC (19.8 mg, 96.0 μmol, 1.50 equiv). It was evacuated and backfilled with nitrogen two times. Then, dichloromethane (500 μL, abs.) was added, and the reaction mixture was stirred for 30 min. A separate vial with DMAP (7.82 mg, 64.0 μmol, 1.00 equiv) and (−)-menthol (10.0 mg, 64.0 μmol, 1.00 equiv) was sealed, evacuated, and backfilled with nitrogen two times before dichloromethane (500 μL, abs.) was added while stirring. After 30 min, the latter solution was added dropwise to the carboxylic acid-containing solution. The reaction was then stirred for 47 h at room temperature. Subsequently, the mixture was poured into an excess of water and extracted with dichloromethane three times. The combined organic extracts were washed with brine, dried over Na_2SO_4, and concentrated under reduced pressure. The crude mixture was purified employing flash chromatography using a hexane/ethyl acetate gradient. The product (8.00 mg, 17.6 μmol, 28% yield) was obtained as an orange-red oil.

^{1}H NMR (400 MHz, CDCl$_3$, (*E*)-isomer) δ = 8.13 (dd, ^3J = 8.8 Hz, ^4J = 2.0 Hz, 2H, 2'-H), 7.86–7.83 (m, 2H, 3'-H), 6.24 (s, 2H, 7'-H), 4.95 (td, J = 10.8 Hz, J = 4.3 Hz, 1H, 1-H), 3.90 (s, 6H, OCH_3), 3.89 (s, 3H, OCH_3), 2.19–2.12 (m, 1H, 6-H), 1.98 (dtd, J = 14.5 Hz, J = 7.2 Hz, J = 3.2 Hz, 1H, 7-H), 1.77–1.71 (m, 2H, 3-H, 4-H), 1.62–1.52 (m, 2H, 2-H, 5-H), 1.17–1.10 (m, 2H, 3-H, 6-H), 0.93 (dd, J = 6.8 Hz, J = 4.2 Hz, 7H, CH_3, C(CH_3)$_2$, 4-H), 0.81 (d, J = 6.9 Hz, 3H, CH_3, C(CH_3)$_2$) ppm.

^{13}C NMR (101 MHz, CDCl$_3$, (*E*)-isomer) δ = 165.97 (C=O), 163.08 (C-8'), 156.89 (C-4'), 155.73 (2C, C-6'), 131.47 (C-1'), 130.58 (+, 2C, C-2'), 127.86 (C-5'), 122.10 (+, 2C, C-3'), 91.60 (+, 2C, C-7'), 75.15 (+, C-1), 56.65 (+, 2C, OCH_3), 55.70 (+, OCH_3), 47.43 (+, C-2), 41.13 (–, C-6), 34.49 (–, C-4), 31.62 (+, C-5), 26.70 (+, C-7), 23.85 (–, C-3), 22.20 (+, CH_3), 20.91 (+, C(CH_3)$_2$), 16.73 (+, C(CH_3)$_2$) ppm.

IR (ATR, ṽ) = 2953 (m), 2927 (m), 2867 (w), 1707 (vs), 1595 (vs), 1575 (vs), 1455 (s), 1438 (s), 1414 (s), 1332 (s), 1283 (s), 1266 (vs), 1230 (vs), 1205 (vs), 1188 (s), 1145 (vs), 1122 (vs), 1111 (vs), 1092 (vs), 1058 (s), 1033 (s), 1010 (s), 982 (m), 958 (s), 915 (m), 863 (m), 844 (w), 812 (s), 773 (vs), 731 (m), 698 (m), 639 (w), 552 (w), 526 (w), 465 (m), 411 (w) cm^{-1}.

MS (EI, 70 eV, 140 °C), m/z (%): 454 (86) [M+H]$^+$, 288 (39) [M-C$_9$H$_{11}$O$_3$+H]$^+$, 195 (100) [C$_9$H$_{11}$N$_2$O$_3$]$^+$.

HRMS–EI *(m/z)*: Calc. for [C$_{26}$H$_{34}$O$_5$N$_2$]$^+$: 454.2462; found: 454.2461.

R$_f$ = 0.13 (Cyclohexane : ethyl acetate = 4:1).

(1*R*,2*S*,5*R*)-2-isopropyl-5-methylcyclohexyl 4-((*E*)-(3,4,5-trimethoxyphenyl)diazenyl) benzoate (84b)

A flask was charged with 4-((3,4,5-trimethoxyphenyl)diazenyl)benzoic acid (30.4 mg, 96.0 μmol, 1.00 equiv) and DCC (29.7 mg, 144 μmol, 1.50 equiv). It was evacuated and backfilled with nitrogen two times. Then, dichloromethane (1.00 mL, abs.) was added, and the reaction mixture was stirred for 30 min. A separate vial with DMAP (11.7 mg, 96.0 μmol, 1.00 equiv) and (−)-menthol (15.0 mg, 96.0 μmol, 1.00 equiv) was sealed, evacuated, and backfilled with nitrogen two times before dichloromethane (1.00 mL, abs.) was added while stirring. After 30 min, the latter solution was added dropwise to the carboxylic acid-containing solution. The reaction was then stirred for 17 h at room temperature. Subsequently, the mixture was poured into an excess of water and extracted with dichloromethane three times. The combined organic extracts were washed with brine, dried over Na_2SO_4, and concentrated under reduced pressure. The crude mixture was purified employing flash chromatography using a hexane/ethyl acetate gradient. The product (21.0 mg, 46.2 μmol, 48% yield) was obtained as an orange-red oil.

^{1}H NMR (400 MHz, CDCl$_3$) δ = 8.18 (dd, ^3J = 8.5 Hz, ^4J = 1.5 Hz, 2H, 2'-H), 7.98–7.87 (m, 2H, 3'-H), 7.29 (d, ^4J = 1.4 Hz, 2H, 6'-H), 4.97 (td, J = 10.9 Hz, J = 4.3 Hz, 1H, 1-H), 3.98 (d, J = 1.4 Hz, 6H, O*CH$_3$*), 3.95 (d, J = 1.4 Hz, 3H, O*CH$_3$*), 2.20–2.11 (m, 1H, 6-H), 1.98 (pd, J = 6.9 Hz, J = 2.3 Hz, 1H, 7-H), 1.74 (dt, J = 14.7 Hz, J = 4.0 Hz, 2H, 3-H, 4-H), 1.58 (ddt, J = 14.0 Hz, J = 6.2 Hz, J = 3.2 Hz, 2H, 5-H, 2-H), 1.15 (dtd, J = 15.3 Hz, J = 11.9 Hz, J = 11.3 Hz, J = 7.4 Hz, 2H, 3-H, 6-H), 0.94 (dt, J = 6.9 Hz, J = 1.9 Hz, 7H, C*H$_3$*, CH(C*H$_3$*)$_2$, 4-H), 0.82 (dd, J = 7.0 Hz, J = 1.3 Hz, 3H, CH(C*H$_3$*)$_2$) ppm.

^{13}C NMR (101 MHz, CDCl$_3$) δ = 165.67 (C=O), 155.12 (C-4'), 153.69 (2C, C-7'), 148.55 (C-5'), 141.43 (C-8'), 132.49 (C-1'), 130.71 (+, 2C, C-2'), 122.60 (+, 2C, C-3'), 100.93 (+, 2C, C-6'), 75.35 (+, C-1), 61.21 (+, O*CH$_3$*), 56.38 (+, 2C, O*CH$_3$*), 47.41 (+, C-2), 41.12 (−, C-6), 34.44 (−, C-4), 31.60 (+, C-5), 26.69 (+, C-7), 23.80 (−, C-3), 22.18 (+, *CH$_3$*), 20.91 (+, CH(*CH$_3$*)$_2$), 16.69 (+, CH(*CH$_3$*)$_2$) ppm.

MS (EI, 70 eV, 120 °C), m/z (%): 454 (100) [M]$^+$, 167 (54) [C$_9$H$_{11}$O$_3$]$^+$.

HRMS–EI *(m/z)*: Calc. for $[C_{26}H_{34}O_5N_2]^+$: 454.2462; found: 454.2462.

IR (ATR, $\tilde{v}$) = 2953 (m), 2932 (m), 2868 (w), 1710 (vs), 1596 (m), 1581 (w), 1494 (s), 1466 (s), 1453 (s), 1429 (w), 1412 (s), 1330 (m), 1306 (s), 1283 (s), 1266 (vs), 1239 (vs), 1220 (vs), 1180 (m), 1152 (w), 1115 (vs), 1094 (vs), 1060 (w), 1038 (m), 1006 (vs), 982 (s), 960 (s), 915 (m), 894 (w), 863 (s), 843 (s), 783 (w), 773 (s), 722 (m), 697 (s), 676 (w), 637 (m), 615 (w), 509 (w), 482 (w), 463 (w), 414 (w), 399 (w) cm^{-1}.

$\mathbf{R}_f$ = 0.50 (Cyclohexane : ethyl acetate = 4:1).

5.2.5. Photoswitchable Fluorophore Moelcule Design

(*E*)-1-(4-bromophenyl)-2-(4-fluorophenyl)diazene (86)

A solution of 4-bromoaniline (500 mg, 333 µL, 2.91 mmol, 1.00 equiv.) in dichloromethane (10.0 mL) was stirred vigorously. At room temperature, a solution of oxone (3.06 g, 4.85 mmol, 1.67 equiv.) in water (10.0 mL) was added dropwise. After 3 h, the organic phase was separated and washed with water, dried over Na_2SO_4 and concentrated under reduced pressure in a flask. 4-Fluoroaniline (323 mg, 279 µL, 2.91 mmol, 1.00 equiv.) was added to this flask and the atmosphere was exchanged for a nitrogen atmosphere. Then, acetic acid (10.0 mL) was added, and the reaction mixture was stirred for 4 d at room temperature. After completion of the reaction, all volatile components were removed under reduced pressure. The product (177 mg, 634 µmol, 22% yield) was obtained by recrystallization from a dichloromethane/methanol mixture as an orange crystalline solid.

^{1}H NMR (400 MHz, CDCl$_3$) δ = 7.94 (dd, $^3J_{H,H}$ = 8.6 Hz, $^4J_{H,F}$ = 5.2 Hz, 2H, 3-H), 7.78 (d, $^3J_{H,H}$ = 8.4 Hz, 2H, 6-H), 7.64 (d, $^3J_{H,H}$ = 8.4 Hz, 2H, 7-H), 7.20 (t, $^3J_{H,H}$ = 8.6 Hz, 2H, 2-H) ppm.

^{13}C NMR (101 MHz, CDCl$_3$) δ = 164.68 (d, $^1J_{C,F}$ = 252.55 Hz, C-1), 151.28 (C-5), 149.11 (d, $^4J_{C,F}$ = 3.02 Hz, C-4), 132.48 (2C, C-6), 125.55 (C-8), 125.13 (d, $^3J_{C,F}$ = 8.8 Hz, 2C, C-3), 124.45 (2C, C-7), 116.26 (d, $^2J_{C,F}$ = 22.9 Hz, 2C, C-2) ppm.

MS (FAB, 3-NBA), m/z (%): 280 (26) [M(^{81}Br)+H]$^+$, 279 (23) [M(^{81}Br)]$^+$, 278 (22) [M(^{79}Br)+H]$^+$.

HRMS–FAB *(m/z)*: Calc. for [C$_{12}$H$_8$N$_2$^{79}BrF]$^+$: 277.9849; found: 277.9847.

IR (ATR, ṽ) = 1901 (w), 1656 (w), 1589 (m), 1572 (m), 1497 (s), 1476 (s), 1408 (m), 1394 (m), 1299 (w), 1227 (s), 1147 (m), 1136 (m), 1095 (m), 1064 (s), 1004 (s), 834 (vs), 779 (m), 715 (s), 664 (s), 637 (m), 538 (vs), 520 (s), 408 (s), 394 (m) cm^{-1}.

EA: Calc. for C$_{12}$H$_8$BrFN$_2$: C 51.64; H 2.89; N 10.04. Found: C 51.43; H 2.80; N 9.94%.

R$_f$ = 0.77 (Cyclohexane : ethyl acetate = 4:1).

228

(*E*)-9-(4-((4-bromophenyl)diazenyl)phenyl)-9*H*-carbazole (87)

A sealable reaction vial was charged with 1-(4-bromophenyl)-2-(4-fluorophenyl)diazene (50.0 mg, 179 µmol, 1.00 equiv.), 9*H*-Carbazole (30.0 mg, 179 µmol, 1.00 equiv.) and K_2CO_3 (49.5 mg, 358 µmol, 2.00 equiv.). It was then sealed and evacuated, and backfilled with nitrogen. Then, DMSO (500 µL, abs.) was added, and the reaction mixture was stirred at 150 °C for 12 h. Subsequently, the product mixture was poured into an excess of water and extracted with dichloromethane (3×). The combined organic phases were washed with brine and dried over Na_2SO_4. After filtration and concentration under reduced pressure, the product (76.0 mg, 178 µmol, 100% yield) was obtained as an orange solid.

¹H NMR (400 MHz, CDCl₃) δ = 8.20–8.14 (m, 4H, 6-H, 13-H), 7.87 (d, ^3J = 8.3 Hz, 2H, 3-H), 7.76 (d, ^3J = 8.2 Hz, 2H, 7-H), 7.70 (d, ^3J = 8.3 Hz, 2H, 2-H), 7.52 (d, ^3J = 8.3 Hz, 2H, 10-H), 7.45 (t, ^3J = 7.7 Hz, 2H, 11-H), 7.33 (t, ^3J = 7.4 Hz, 2H, 12-H) ppm.

¹³C NMR (101 MHz, CDCl₃) δ = 151.48 (C-4), 151.09 (C-5), 140.59 (2C, C-9), 140.55 (C-8), 132.59 (2C, C-2), 127.45 (2C, C-7), 126.31 (2C, C-11), 125.84 (C-1), 124.65 (2C, C-6), 124.61 (2C, C-3), 123.89 (2C, C-14), 120.59 (4C, C-12, C-13), 109.99 (2C, C-10) ppm.

IR (ATR, ṽ) = 3044 (w), 2919 (m), 2850 (m), 1596 (m), 1568 (w), 1503 (m), 1479 (m), 1442 (s), 1417 (m), 1394 (m), 1360 (m), 1340 (m), 1332 (m), 1312 (s), 1299 (m), 1282 (w), 1220 (vs), 1187 (m), 1177 (m), 1153 (m), 1142 (m), 1125 (m), 1103 (m), 1064 (m), 1028 (w), 1001 (m), 912 (m), 851 (s), 833 (vs), 734 (vs), 713 (vs), 703 (vs), 679 (m), 646 (m), 620 (s), 598 (m), 562 (m), 544 (vs), 528 (s), 426 (m), 416 (s), 405 (m) cm⁻¹.

R$_f$ = 0.85 (Cyclohexane : ethyl acetate = 2:1).

(*E*)-4'-((4-(9*H*-carbazol-9-yl)phenyl)diazenyl)-[1,1'-biphenyl]-4-carbonitrile (88)

To a solution of (*E*)-9-(4-((4-bromophenyl)diazenyl)phenyl)-9*H*-carbazole (18.0 mg, 42.2 μmol, 1.00 equiv.) in 1,4-dioxane (500 μL) and water (100 μL), cesium carbonate (55.0 mg, 169 μmol, 4.00 equiv.) was added while stirring. Then, [Pd(PPh₃)₂Cl₂] (5.93 mg, 8.44 μmol, 0.20 equiv.) and (4-cyanophenyl)boronic acid (9.31 mg, 63.3 μmol, 1.50 equiv.) were added and the reaction mixture was heated at 110 °C for 15 h. Subsequently, the reaction mixture was poured into an excess of water and extracted three times with dichloromethane. The combined organic phases were washed with brine and dried over Na_2SO_4 before they were concentrated under reduced pressure. The crude mixture was further purified employing column chromatography (hexane : ethyl acetate = 100:1 → 50:1 → 20:1 → 10:1). The product (2.00 mg, 4.46 μmol, 11% yield) was obtained as a yellow crystalline solid.

¹H NMR (400 MHz, CDCl₃) δ = 8.21 (d, ^3J = 8.6 Hz, 2H, 7-H), 8.17 (d, ^3J = 7.7 Hz, 2H, 17-H), 8.10 (d, ^3J = 8.5 Hz, 2H, 10-H), 7.83–7.75 (m, 8H, 2-H, 3-H, 6-H, 11-H), 7.53 (d, ^3J = 8.2 Hz, 2H, 14-H), 7.49–7.42 (m, 2H, 15-H), 7.36–7.30 (m, 2H, 16-H) ppm.

¹³C NMR (101 MHz, CDCl₃) δ = 152.53 (C-9), 151.13 (C-8), 144.61 (C-4), 141.69 (C-12), 140.47 (2C, C-13), 140.45 (C-5), 132.76 (2C, C-2), 128.12 (2C, C-6), 127.84 (2C, C-3), 127.33 (2C, C-11), 126.18 (2C, C-15), 124.56 (2C, C-7), 123.77 (2C, C-10), 123.75 (2C, C-18), 120.48 (4C, C-16, C-17), 118.79 (*C*N), 111.59 (C-1), 109.87 (2C, C-14) ppm.

HRMS–FAB *(m/z)*: Calc. for [C₃₁H₂₁N₄]⁺: 449.1716; found: 449.1716.

R$_f$ = 0.51 (Cyclohexane : ethyl acetate = 4:1).

5.3. Crystal Structures

Dr. Martin Nieger (University of Helsinki) measured and solved the following crystal structures.

The single-crystal X-ray diffraction study was conducted on a Bruker D8 Venture diffractometer with a PhotonII detector at 298(2) K or 173(2) K using Cu-Kα radiation ($\lambda =$ 1.54178 Å). Dual space methods (SHELXT)[171] were used for the structure solution, and refinement was done using SHELXL-2014 (full-matrix least-squares on F^2)[172]. Hydrogen atoms were refined using a riding model. Semi-empirical absorption corrections were applied. **2DMAC-BP-F** is refined as an inversion twin. In **2DMAC-BP-F**, the F-atom is disordered about a mirror plane. In **2DTCz-BP-F** are four crystallographic independent molecules in the asymmetric unit. In each molecule, the F-atoms are disordered (about a mirror plane). In two voids are twenty benzene solvent molecules per void; due to the bad quality of the data of **2DTCz-BP-F**, the data were not deposited with The Cambridge Crystallographic Data Centre.

2DMAC-BP-F (sb1439_hy): red crystals, $C_{50}H_{37}FN_4 \cdot 2(CH_2Cl_2)$, $M_r = 882.69$, crystal size 0.36 × 0.12 × 0.06 mm, orthorhombic, space group $Pmn2_1$ (No. 31), $a = 16.2097(7)$ Å, $b = 7.3687(3)$ Å, $c = 18.0747(7)$ Å, $V = 2158.92(15)$) Å^3, $Z = 2$, $\rho = 1.358$ Mg/m^{-3}, μ(Cu-K$_\alpha$) = 2.86 mm^{-1}, $F(000) = 916$, $T = 298$ K, $2\theta_{max} = 145.2°$, 38535 reflections, of which 4395 were independent ($R_{int} = 0.034$), 290 parameters, 255 restraints (see cif-file for details), $R_1 = 0.046$ (for 4246 I > 2σ(I)), w$R_2 = 0.131$ (all data), $S = 1.08$, largest diff. peak / hole = 0.37 / -0.39 e Å^{-3}, (BASF = 0.48(3)).

2DTCz-BP-F (sb1426_hy): yellow crystals, $C_{60}H_{57}FN_4 \cdot 2.5(C_6H_6)$, $M_r = 1048.36$, crystal size 0.20 × 0.16 × 0.06 mm, monoclinic, space group $P2_1/c$ (No. 14), $a = 45.2341(15)$ Å, $b = 14.5591(5)$ Å, $c = 41.0844(14)$ Å, $\beta = 113.749(2)°$, $V = 24765.6(15)$ Å^3, $Z = 16$, $\rho = 1.125$ Mg/m^{-3}, $\mu(Cu\text{-}K_\alpha) = 0.52$ mm^{-1}, $F(000) = 8944$, $T = 173$ K, $2\theta_{max} = 144.4°$, 223270 reflections, of which 48596 were independent ($R_{int} = 0.036$) (38367 with I > 2σ(I)).

CCDC 2086119 (**2DMAC-BP-F**) contains the supplementary crystallographic data for these structures. This data can be obtained for free from The Cambridge Crystallographic Data Centre via www.ccdc.cam.ac.uk/data_request/cif. Due to the bad quality of the data of **2DTCz-BP-F**, the data were not deposited with The Cambridge Crystallographic Data Centre.

6. Abbreviations

3-NBA	3-Nitrobenzyl alcohol
A	Acceptor
AB	Azobenzene
AIDF	Aggregation-induced delayed fluorescence
AntPhos	4-(Anthracen-9-yl)-3-(*tert*-butyl)-2,3-dihydrobenzo[*d*][1,3]oxaphosphole
aq.	Aqueous
atm	Standard atmosphere
ATR	Attenuated total reflection
BP	Dibenzo[*a,c*]phenazine
calc.	Calculated
CBP	4,4'-bis-(*N*-carbazolyl)-1,1'-biphenyl
CE_{max}	Maximal current efficiency
CIE	Commission internationale de l'éclairage
COPh	Benzoyl
CT	Charge transfer
CV	Cyclic voltammetry
Cy	Cyclohexyl
D	Donor
dba	Dibenzylideneacetone
DCM	Dichloromethane
DEPT	Distortionless enhancement by polarization transfer
DF	Delayed Fluorescence
DFT	Density functional theory
DMAC	9,9-Dimethyl-9,10-dihydroacridine
DMAP	4-Dimethylaminopyridine
DMF	*N,N*-Dimethylformamide
DMSO	Dimethyl sulfoxide
Do	Donor
dppf	1,1'-Bis(diphenylphosphino)ferrocene
DRG	Dorsal root ganglion
DTCz	3,6-Di-*tert*-butyl-carbazole
EA	Elemental Analysis
EBL	Electron blocking layer
EDC	(3-Dimethylaminopropyl)-3-ethylcarbodiimide

EI	Electron impact
EI	Electron ionization
EL	Electroluminescence
EML	Emissive layer
EQE	External quantum efficiency
equiv.	Equivalent
ESI	Electrospray ionization
ETL	Electron transport layer
eV	Electronvolt
EWG	Electron withdrawing group
f	Oscillator strength
F	Fluorescence
FAB	Fast atom bombardment
Fc	Ferrocene
FMO	Frontier molecular orbitals
GC	Glassy carbon
HATCN	1,4,5,8,9,11-Hexaazatriphenylene- hexacarbonitrile
HBL	Hole blocking layer
HBL	Hole blocking layer
HIL	Hole injection layer
HOMO	Highest occupied molecular orbital
HRMS	High-resolution mass spectrometry
HSQC	Heteronuclear single quantum coherence
HTL	Hole transport layer
Hz	Hertz
IC	Internal conversion
IC50	Half maximal inhibitory concentration
ICT	Intramolecular charge transfer
IQE	Internal quantum efficiency
IR	Infrared
ISC	Intersystem crossing
ITO	Indium tin oxide $((In2O3)0.9(SnO2)0.1)$
lm	Lumen
LUMO	Lowest unoccupied molecular orbital
M	Molar
m.p.	Melting point

MAL	Maleimide
Me	Methyl
Mes	Mesityl
MOF	Metal-organic framework
MTS	3-(4,5-Dimethylthiazol-2-yl)-5-(3-carboxymethoxyphenyl)-2-(4-sulfophenyl)-2H-tetrazolium
MTT	3-(4,5-Dimethylthiazol-2-yl)-2,5-diphenyltetrazolium bromide
nBu	n-Butyl
nBuLi	n-Butyllithium
nBuOH	n-Butanol
NID	6-(9,9-Dimethylacridin-10(9H)-yl)-2-phenyl-1H-benzo[de]isoquinoline-1,3(2H)-dione
NMR	Nuclear magnetic resonance
NPB	N,N'-Di(1-naphthyl)- N,N'-diphenyl-(1,1'-biphenyl)-4,4'-diamine
OLED	Organic light-emitting diode
OMe	Methoxy
OtBu	$tert$-Butoxide
PBS	Phosphate-buffered saline
PDT	Photodynamic therapy
PE$_{max}$	Maximum power efficiency
PF	Prompt Fluorescence
Ph	Phenyl
PhMe	Toluene
PIT	Pyrrolo[3,4-f]isoindole-1,3,5,7(2H,6H)-tetraone
PL	Photoluminescence
PLQY	Photoluminescence quantum yield
PMMA	Poly(methyl methacrylate)
ppm	Parts per million
PQD	1H-Pyrrolo[3,4-b]quinoxaline-1,3(2H)-dione
PS	Photosensitizer
PSS	Photostationary state
pTsOH	$para$-Toluenesulfonic acid
PXZ	10H-phenoxazine
QA	Quaternary amine
quant.	Quantitative
r.t.	Room temperature

236

R$_f$	Retardation factor
RISC	Reverse intersystem crossing
ROS	Reactive oxygen species
rRNA	Ribosomal ribonucleic acid
RSA	Rhodamine spiroamide
rTRPM8	Rat TRPM8 orthologue
sat.	Saturated
SBU	Secondary building units
SCE	Saturated calomel electrode
SNAr	Nucleophilic aromatic substitution
SOC	Spin-orbit coupling
SPARK	Synthetic photoisomerizable azobenzene-regulated K+
SURMOF	Surface-anchored MOF
TADF	Thermally activated delayed fluorescence
TBAF	Tetrabutylammonium fluoride
tBu	$tert$-Butyl
TCTA	Tris(4-carbazoyl-9-ylphenyl)amine
TD-DFT	Time-dependent DFT
TDA	TAMM-DANCOFF approximation
THF	Tetrahydrofuran
TLC	Thin layer chromatography
TmPyPB	1,3,5-Tri[(3-pyridyl)-phen-3-yl]benzene
TMS	Trimethylsilyl
TPA	Triphenylphosphine
TPP	Triphenylphosphonium
TRLI	Time-resolved luminescence imaging
TRP	Transient receptor protein
TRPM8	Transient receptor potential melastatin subtype 8
TRPV1	Transient receptor potential cation channel subfamily V member 1
TSCT	Through-space charge transfer
UV	Ultraviolet
vis	Visible
VR	Vibrational relaxation
δ	Chemical shift
ε	Extinction coefficient

7. References

[1] Y. Liu, C. Li, Z. Ren, S. Yan, M. R. Bryce, *Nat. Rev. Mater.* **2018**, *3*, 18020.

[2] G. Hong, X. Gan, C. Leonhardt, Z. Zhang, J. Seibert, J. M. Busch, S. Bräse, *Adv. Mater.* **2021**, *33*, e2005630.

[3] F. Fang, L. Zhu, M. Li, Y. Song, M. Sun, D. Zhao, J. Zhang, *Adv. Sci.* **2021**, *8*, e2102970.

[4] M. J. Fuchter, *J. Med. Chem.* **2020**, *63*, 11436–11447.

[5] K. Hull, J. Morstein, D. Trauner, *Chem. Rev.* **2018**, *118*, 10710–10747.

[6] C. Bizzarri, E. Spuling, D. M. Knoll, D. Volz, S. Bräse, *Coord. Chem. Rev.* **2018**, *373*, 49–82.

[7] X. Xiong, F. Song, J. Wang, Y. Zhang, Y. Xue, L. Sun, N. Jiang, P. Gao, L. Tian, X. Peng, *J. Am. Chem. Soc.* **2014**, *136*, 9590–9597.

[8] F. Ni, N. Li, L. Zhan, C. Yang, *Adv. Opt. Mater.* **2020**, *8*, 1902187.

[9] D. E. Dolmans, D. Fukumura, R. K. Jain, *Nat. Rev. Cancer* **2003**, *3*, 380–387.

[10] I. S. Park, S. Y. Lee, C. Adachi, T. Yasuda, *Adv. Funct. Mater.* **2016**, *26*, 1813–1821.

[11] J. Li, K. Pu, *Chem. Soc. Rev.* **2019**, *48*, 38–71.

[12] A. P. Castano, P. Mroz, M. R. Hamblin, *Nat. Rev. Cancer* **2006**, *6*, 535–545.

[13] X. Cui, J. Zhang, Y. Wan, F. Fang, R. Chen, D. Shen, Z. Huang, S. Tian, Y. Xiao, X. Li, J. Chelora, Y. Liu, W. Zhang, C. S. Lee, *ACS Appl. Bio. Mater.* **2019**, *2*, 3854–3860.

[14] N. L. Oleinick, R. L. Morris, I. Belichenko, *Photochem. Photobiol. Sci.* **2002**, *1*, 1–21.

[15] W. Fan, P. Huang, X. Chen, *Chem. Soc. Rev.* **2016**, *45*, 6488–6519.

[16] B. Valeur, *Molecular Fluorescence*, Wiley-VCH, Weinheim, Germany, **2001**.

[17] N. J. Turro, V. Ramanmurthy, J. C. Scaiano, *Principles of Molecular Photochemistry: An Introduction*, University Science Books, **2009**.

[18] V. Balzani, P. Ceroni, A. Juris, *Photochemistry and Photophysics*, Wiley-VCH, Weinheim, Germany, **2014**.

[19] R. Ishimatsu, S. Matsunami, K. Shizu, C. Adachi, K. Nakano, T. Imato, *J. Phys. Chem. A* **2013**, *117*, 5607–5612.

[20] F. B. Dias, T. J. Penfold, A. P. Monkman, *Methods Appl. Fluoresc.* **2017**, *5*, 012001.

238

[21]   B. Milián-Medina, J. Gierschner, *Org. Electron.* **2012**, *13*, 985–991.

[22]   P. K. Samanta, D. Kim, V. Coropceanu, J. L. Bredas, *J. Am. Chem. Soc.* **2017**, *139*, 4042–4051.

[23]   Z. Yang, Z. Mao, C. Xu, X. Chen, J. Zhao, Z. Yang, Y. Zhang, W. Wu, S. Jiao, Y. Liu, M. P. Aldred, Z. Chi, *Chem. Sci.* **2019**, *10*, 8129–8134.

[24]   E. Spuling, N. Sharma, I. D. W. Samuel, E. Zysman-Colman, S. Bräse, *Chem. Commun.* **2018**, *54*, 9278–9281.

[25]   X. Wang, S. Wang, J. Lv, S. Shao, L. Wang, X. Jing, F. Wang, *Chem. Sci.* **2019**, *10*, 2915–2923.

[26]   J. Hu, Q. Li, X. Wang, S. Shao, L. Wang, X. Jing, F. Wang, *Angew. Chem. Int. Ed.* **2019**, *58*, 8405–8409.

[27]   P. de Silva, C. A. Kim, T. Zhu, T. Van Voorhis, *Chem. Mater.* **2019**, *31*, 6995–7006.

[28]   Q. Zhang, B. Li, S. Huang, H. Nomura, H. Tanaka, C. Adachi, *Nat. Photonics* **2014**, *8*, 326–332.

[29]   J.-X. Chen, K. Wang, C.-J. Zheng, M. Zhang, Y.-Z. Shi, S.-L. Tao, H. Lin, W. Liu, W.-W. Tao, X.-M. Ou, X.-H. Zhang, *Adv. Sci.* **2018**, *5*, 1800436.

[30]   H. Tanaka, K. Shizu, H. Miyazaki, C. Adachi, *Chem. Commun.* **2012**, *48*, 11392–11394.

[31]   M. Segal, M. A. Baldo, R. J. Holmes, S. R. Forrest, Z. G. Soos, *Phys. Rev. B* **2003**, *68*.

[32]   T. J. Penfold, *J. Phys. Chem. C* **2015**, *119*, 13535–13544.

[33]   A. Dreuw, M. Head-Gordon, *Chem. Rev.* **2005**, *105*, 4009–4037.

[34]   M. Moral, L. Muccioli, W. J. Son, Y. Olivier, J. C. Sancho-Garcia, *J. Chem. Theory Comput.* **2015**, *11*, 168–177.

[35]   H. Uoyama, K. Goushi, K. Shizu, H. Nomura, C. Adachi, *Nature* **2012**, *492*, 234–238.

[36]   J. H. Kim, J. H. Yun, J. Y. Lee, *Adv. Opt. Mater.* **2018**, *6*, 1800255.

[37]   R. Englman, J. Jortner, *Mol. Phys.* **1970**, *18*, 145–164.

[38]   J. V. Caspar, E. M. Kober, B. P. Sullivan, T. J. Meyer, *J. Am. Chem. Soc.* **1982**, *104*, 630–632.

[39]   J. Xue, Q. Liang, R. Wang, J. Hou, W. Li, Q. Peng, Z. Shuai, J. Qiao, *Adv. Mater.* **2019**, *31*, e1808242.

[40]   G. Hong, C. Si, A. K. Gupta, C. Bizzarri, M. Nieger, I. D. W. Samuel, E. Zysman-Colman, S. Bräse, *J. Mater. Chem. C* **2022**, *10*, 4757.

239

[41] S. Kothavale, W. J. Chung, J. Y. Lee, *ACS Appl. Mater. Interfaces* **2020**, *12*, 18730–18738.

[42] F. M. Xie, P. Wu, S. J. Zou, Y. Q. Li, T. Cheng, M. Xie, J. X. Tang, X. Zhao, *Adv. Electron. Mater.* **2019**, *6*.

[43] F. M. Xie, H. Z. Li, G. L. Dai, Y. Q. Li, T. Cheng, M. Xie, J. X. Tang, X. Zhao, *ACS Appl. Mater. Interfaces* **2019**, *11*, 26144–26151.

[44] Y. Liu, Y. Chen, H. Li, S. Wang, X. Wu, H. Tong, L. Wang, *ACS Appl. Mater. Interfaces* **2020**, *12*, 30652–30658.

[45] W. Zeng, T. Zhou, W. Ning, C. Zhong, J. He, S. Gong, G. Xie, C. Yang, *Adv. Mater.* **2019**, *31*, e1901404.

[46] Y. Hu, Y. Zhang, W. Han, J. Li, X. Pu, D. Wu, Z. Bin, J. You, *Chem. Eng. J.* **2022**, *428*, 131186.

[47] Q. Zhang, S. Xu, M. Li, Y. Wang, N. Zhang, Y. Guan, M. Chen, C. F. Chen, H. Y. Hu, *Chem. Commun.* **2019**, *55*, 5639–5642.

[48] F. Ni, Z. Zhu, X. Tong, W. Zeng, K. An, D. Wei, S. Gong, Q. Zhao, X. Zhou, C. Yang, *Adv. Sci.* **2019**, *6*, 1801729.

[49] S. Qi, S. Kim, V. N. Nguyen, Y. Kim, G. Niu, G. Kim, S. J. Kim, S. Park, J. Yoon, *ACS Appl. Mater. Interfaces* **2020**, *12*, 51293–51301.

[50] S. Xu, Q. Zhang, X. Han, Y. Wang, X. Wang, M. Nazare, J. D. Jiang, H. Y. Hu, *ACS Sens.* **2020**, *5*, 1650–1656.

[51] R. Wei, L. Zhang, S. Xu, Q. Zhang, Y. Qi, H. Y. Hu, *Chem. Commun.* **2020**, *56*, 2550–2553.

[52] H. Mieno, R. Kabe, M. D. Allendorf, C. Adachi, *Chem. Commun.* **2018**, *54*, 631–634.

[53] H. C. Zhou, S. Kitagawa, *Chem. Soc. Rev.* **2014**, *43*, 5415–5418.

[54] Z. Wei, Z. Y. Gu, R. K. Arvapally, Y. P. Chen, R. N. McDougald, Jr., J. F. Ivy, A. A. Yakovenko, D. Feng, M. A. Omary, H. C. Zhou, *J. Am. Chem. Soc.* **2014**, *136*, 8269–8276.

[55] R. Haldar, M. Jakoby, M. Kozlowska, M. Rahman Khan, H. Chen, Y. Pramudya, B. S. Richards, L. Heinke, W. Wenzel, F. Odobel, S. Diring, I. A. Howard, U. Lemmer, C. Woll, *Chem. Eur. J.* **2020**, *26*, 17016–17020.

[56] J. Yu, Y. Cui, H. Xu, Y. Yang, Z. Wang, B. Chen, G. Qian, *Nat. Commun.* **2013**, *4*, 2719.

[57] B. Bie, L. Guo, M. Zhang, Y. Ma, C. Yang, *Adv. Opt. Mater.* **2021**, *10*, 2101992.

240

[58] Nobelförsamlingen - The Nobel Assembly at Karolinska Institutet, "The Nobel Prize in Physiology or Medicine 2021", can be found under https://www.nobelprize.org/uploads/2021/10/press-medicineprize2021.pdf, **2021**.

[59] D. D. McKemy, W. M. Neuhausser, D. Julius, *Nature* **2002**, *416*, 52−58.

[60] A. M. Peier, A. Moqrich, A. C. Hergarden, A. J. Reeve, D. A. Andersson, G. M. Story, T. J. Earley, I. Dragoni, P. McIntyre, S. Bevan, A. Patapoutian, *Cell* **2002**, *108*, 705−715.

[61] S. S. Bharate, S. B. Bharate, *ACS Chem. Neurosci.* **2012**, *3*, 248−267.

[62] T. Voets, G. Owsianik, B. Nilius, in *Transient Receptor Potential (TRP) Channels. Handbook of Experimental Pharmacology, Vol. 179* (Eds.: V. Flockerzi, B. Nilius), Springer, Berlin, Heidelberg, **2007**, pp. 329-344.

[63] C. Izquierdo, M. Martin-Martinez, I. Gomez-Monterrey, R. Gonzalez-Muniz, *Int. J. Mol. Sci.* **2021**, *22*.

[64] L. Tsavaler, M. H. Shapero, S. Morkowski, R. Laus, *Cancer Res.* **2001**, *61*, 3760−3769.

[65] G. Yosipovitch, C. Szolar, X. Y. Hui, H. Maibach, *Arch. Dermatol. Res.* **1996**, *288*, 245−248.

[66] A. O. Barel, M. Paye, H. I. Maibach, *Handbook of Cosmetic Science and Technology*, 3rd ed. ed., Informa Healthcare USA, Inc., New York, **2009**.

[67] Y. Yin, S. Y. Lee, *Trends Biochem. Sci.* **2020**, *45*, 806−819.

[68] J. A. Farco, O. Grundmann, *Mini Reviews in Medicinal Chemistry* **2013**, *13*, 124−131.

[69] L. Xu, Y. Han, X. Chen, A. Aierken, H. Wen, W. Zheng, H. Wang, X. Lu, Z. Zhao, C. Ma, P. Liang, W. Yang, S. Yang, F. Yang, *Nat. Commun.* **2020**, *11*, 3790.

[70] T. Voets, K. Talavera, G. Owsianik, B. Nilius, *Nat. Chem. Biol.* **2005**, *1*, 85−92.

[71] T. Voets, G. Droogmans, U. Wissenbach, A. Janssens, V. Flockerzi, B. Nilius, *Nature* **2004**, *430*, 748−754.

[72] K. Tsuzuki, H. Xing, J. Ling, J. G. Gu, *J. Neurosci.* **2004**, *24*, 762−771.

[73] M. A. Sherkheli, A. K. Vogt-Eisele, D. Bura, L. R. Beltran Marques, G. Gisselmann, H. Hatt, *J. Pharm. Pharm. Sci.* **2010**, *13*, 242−253.

[74] M. A. Sherkheli, G. Gisselmann, A. K. Vogt-Eisele, J. F. Doerner, H. Hatt, *Pak. J. Pharm. Sci.* **2008**, *21*, 370−378.

[75] L. Cole, S. M. Furrer, C. Galopin, P. V. Krawec, A. Masur, J. P. Slack (Givaudan SA), WO2006092074A1, **2005**.

[76]   V. B. Journigan, D. Alarcon-Alarcon, Z. Feng, Y. Wang, T. Liang, D. C. Dawley, A. Amin, C. Montano, W. D. Van Horn, X. Q. Xie, A. Ferrer-Montiel, A. Fernandez-Carvajal, *ACS Med. Chem. Lett.* **2021**, *12*, 758−767.

[77]   G. Ortar, L. De Petrocellis, L. Morera, A. S. Moriello, P. Orlando, E. Morera, M. Nalli, V. Di Marzo, *Bioorg. Med. Chem. Lett.* **2010**, *20*, 2729−2732.

[78]   S. Crespi, N. A. Simeth, B. König, *Nat. Rev. Chem.* **2019**, *3*, 133−146.

[79]   G. S. Hartley, *Nature* **1937**, *140*, 281−281.

[80]   A. A. Beharry, G. A. Woolley, *Chem. Soc. Rev.* **2011**, *40*, 4422−4437.

[81]   F. Aleotti, A. Nenov, L. Salvigni, M. Bonfanti, M. M. El-Tahawy, A. Giunchi, M. Gentile, C. Spallacci, A. Ventimiglia, G. Cirillo, L. Montali, S. Scurti, M. Garavelli, I. Conti, *J. Phys. Chem. A* **2020**, *124*, 9513−9523.

[82]   C. Xu, L. Yu, F. L. Gu, C. Zhu, *Phys. Chem. Chem. Phys.* **2018**, *20*, 23885−23897.

[83]   A. Cembran, F. Bernardi, M. Garavelli, L. Gagliardi, G. Orlandi, *J. Am. Chem. Soc.* **2004**, *126*, 3234−3243.

[84]   J. Casellas, M. J. Bearpark, M. Reguero, *ChemPhysChem* **2016**, *17*, 3068−3079.

[85]   H. M. Bandara, S. C. Burdette, *Chem. Soc. Rev.* **2012**, *41*, 1809−1825.

[86]   I. Conti, M. Garavelli, G. Orlandi, *J. Am. Chem. Soc.* **2008**, *130*, 5216−5230.

[87]   E. Wei-Guang Diau, *J. Phys. Chem. A* **2004**, *108*, 950−956.

[88]   E. R. Talaty, J. C. Fargo, *Chem. Commun.* **1967**, *0*, 65−66.

[89]   W. F. Cheong, S. A. Prahl, A. J. Welch, *IEEE J. Quantum Electron.* **1990**, *26*, 2166−2185.

[90]   M. Dong, A. Babalhavaeji, S. Samanta, A. A. Beharry, G. A. Woolley, *Acc. Chem. Res.* **2015**, *48*, 2662−2670.

[91]   J. J. Chambers, M. R. Banghart, D. Trauner, R. H. Kramer, *J. Neurophysiol.* **2006**, *96*, 2792−2796.

[92]   M. Banghart, K. Borges, E. Isacoff, D. Trauner, R. H. Kramer, *Nat. Neurosci.* **2004**, *7*, 1381−1386.

[93]   R. O. Blaustein, P. A. Cole, C. Williams, C. Miller, *Nat. Struct. Biol.* **2000**, *7*, 309−311.

[94]   C. K. McKenzie, I. Sanchez-Romero, H. Janovjak, in *Novel Chemical Tools to Study Ion Channel Biology, Vol. 869* (Eds.: C. Ahern, S. Pless), Springer Science+Business Media, New York, **2015**.

242

[95]   M. Stein, A. Breit, T. Fehrentz, T. Gudermann, D. Trauner, *Angew. Chem. Int. Ed.* **2013**, *52*, 9845–9848.

[96]   J. A. Frank, M. Moroni, R. Moshourab, M. Sumser, G. R. Lewin, D. Trauner, *Nat. Commun.* **2015**, *6*, 7118.

[97]   D. B. Konrad, J. A. Frank, D. Trauner, *Chem. Eur. J.* **2016**, *22*, 4364–4368.

[98]   D. B. Konrad, G. Savasci, L. Allmendinger, D. Trauner, C. Ochsenfeld, A. M. Ali, *J. Am. Chem. Soc.* **2020**, *142*, 6538–6547.

[99]   T. Fukaminato, E. Tateyama, N. Tamaoki, *Chem. Commun.* **2012**, *48*, 10874–10876.

[100] M. Han, D. Ishikawa, E. Muto, M. Hara, *J. Lumin.* **2009**, *129*, 1163–1168.

[101] M. Shimomura, T. Kunitake, *J. Am. Chem. Soc.* **2002**, *109*, 5175–5183.

[102] M. Han, M. Hara, *J. Am. Chem. Soc.* **2005**, *127*, 10951–10955.

[103] M. R. Han, M. Hara, *New J. Chem.* **2006**, *30*, 223–227.

[104] S. Wang, Y. Miao, X. Yan, K. Ye, Y. Wang, *J. Mater. Chem. C* **2018**, *6*, 6698–6704.

[105] J. X. Chen, W. W. Tao, Y. F. Xiao, K. Wang, M. Zhang, X. C. Fan, W. C. Chen, J. Yu, S. Li, F. X. Geng, X. H. Zhang, C. S. Lee, *ACS Appl. Mater. Interfaces* **2019**, *11*, 29086–29093.

[106] S. Kothavale, W. J. Chung, J. Y. Lee, *J. Mater. Chem. C* **2020**, *8*, 7059–7066.

[107] C. Zhou, W.-C. Chen, H. Liu, X. Cao, N. Li, Y. Zhang, C.-S. Lee, C. Yang, *J. Mater. Chem. C* **2020**, *8*, 9639–9645.

[108] U. Balijapalli, Y. T. Lee, B. S. B. Karunathilaka, G. Tumen-Ulzii, M. Auffray, Y. Tsuchiya, H. Nakanotani, C. Adachi, *Angew. Chem. Int. Ed.* **2021**, *60*, 19364–19373.

[109] K. Sun, Z. Cai, J. Jiang, W. Tian, W. Guo, J. Shao, W. Jiang, Y. Sun, *Dyes Pig.* **2020**, *173*.

[110] J. X. Zhou, X. Y. Zeng, F. M. Xie, Y. H. He, Y. Q. Tang, Y. Q. Li, J. X. Tang, *Mater. Today Energy* **2021**, *21*.

[111] S. Kothavale, J. Lim, J. Yeob Lee, *Chem. Eng. J.* **2022**, *431*, 134216.

[112] J.-L. He, F.-C. Kong, B. Sun, X.-J. Wang, Q.-S. Tian, J. Fan, L.-S. Liao, *Chem. Eng. J.* **2021**, *424*.

[113] V. Andruleviciene, K. Leitonas, D. Volyniuk, G. Sini, J. V. Grazulevicius, V. Getautis, *Chem. Eng. J.* **2021**, *417*.

243

[114]   F. Hundemer, *Modular Design Strategies for TADF Emitters: Towards Highly Efficient Materials for OLED Application*, Logos Verlag, Berlin, **2020**.

[115]   J. Broichhagen, J. A. Frank, D. Trauner, *Acc. Chem. Res.* **2015**, *48*, 1947–1960.

[116]   W. A. Velema, W. Szymanski, B. L. Feringa, *J. Am. Chem. Soc.* **2014**, *136*, 2178–2191.

[117]   J. H. Maeng, R. Braveenth, Y. H. Jung, S. J. Hwang, H. Lee, H. L. Min, J. Y. Kim, C. W. Han, J. H. Kwon, *Dyes Pigm.* **2021**, *194*.

[118]   C. Zhou, S. Xiao, M. Wang, W. Jiang, H. Liu, S. Zhang, B. Yang, *Front. Chem.* **2019**, *7*, 141.

[119]   J. Cao, S. Wang, R. Hua (Shijiazhuang Chengzhi Yonghua Display Material Co., Ltd.), CN106967060, **2017**.

[120]   Y.-Y. Wang, Y.-L. Zhang, K. Tong, L. Ding, J. Fan, L.-S. Liao, *J. Mater. Chem. C* **2019**, *7*, 15301–15307.

[121]   D. G. Congrave, B. H. Drummond, P. J. Conaghan, H. Francis, S. T. E. Jones, C. P. Grey, N. C. Greenham, D. Credgington, H. Bronstein, *J. Am. Chem. Soc.* **2019**, *141*, 18390–18394.

[122]   S. Jeong, Y. Lee, J. K. Kim, D.-J. Jang, J.-I. Hong, *J. Mater. Chem. C* **2018**, *6*, 9049–9054.

[123]   G. Mann, J. F. Hartwig, *J. Am. Chem. Soc.* **1996**, *118*, 13109–13110.

[124]   G. Mann, J. F. Hartwig, *J. Org. Chem.* **1997**, *62*, 5413–5418.

[125]   G. Mann, J. F. Hartwig, *Tetrahedron Lett.* **1997**, *38*, 8005–8008.

[126]   M. Palucki, J. P. Wolfe, S. L. Buchwald, *J. Am. Chem. Soc.* **1996**, *118*, 10333–10334.

[127]   R. A. Widenhoefer, H. A. Zhong, S. L. Buchwald, *J. Am. Chem. Soc.* **1997**, *119*, 6787–6795.

[128]   M. Watanabe, M. Nishiyama, Y. Koie, *Tetrahedron Lett.* **1999**, *40*, 8837–8840.

[129]   Q. Shelby, N. Kataoka, G. Mann, J. Hartwig, *J. Am. Chem. Soc.* **2000**, *122*, 10718–10719.

[130]   N. Kataoka, Q. Shelby, J. P. Stambuli, J. F. Hartwig, *J. Org. Chem.* **2002**, *67*, 5553–5566.

[131]   J. X. Chen, Y. F. Xiao, K. Wang, D. Sun, X. C. Fan, X. Zhang, M. Zhang, Y. Z. Shi, J. Yu, F. X. Geng, C. S. Lee, X. H. Zhang, *Angew. Chem. Int. Ed.* **2021**, *60*, 2478–2484.

[132]   Y. Ma, D. Hu, L. Ying, T. Guo (South China University of Technology), CN113121511, **2021**.

[133] D. Dey, M. K. Sarangi, A. Ray, D. Bhattacharyya, D. K. Maity, *J. Lumin.* **2016**, *173*, 105–112.

[134] Y. Li, F. Xie, J. Tang, J. Zhou, X. Zeng (Soochow University), CN112038494, **2020**.

[135] J. Tang, F. Xie, Y. Li, J. Zhou, X. Zeng (Soochow University), CN112079843, **2020**.

[136] C. Zheng, C. Pu, P. Yin (University of Electronic Science and Technology of China), CN112961148, **2021**.

[137] S. Su, W. Li, K. Liu, W. Li, X. Peng (South China University of Technology), CN113214221, **2021**.

[138] J. Liang, C. Li, Y. Cui, Z. Li, J. Wang, Y. Wang, *J. Mater. Chem. C* **2020**, *8*, 1614–1622.

[139] R. Jiang, X. Wu, H. Liu, J. Guo, D. Zou, Z. Zhao, B. Z. Tang, *Adv. Sci.* **2022**, *9*, e2104435.

[140] B. Tang, Z. Zhao, R. Jiang, H. Liu, X. Wu (South China University of Technology), CN114014841, **2022**.

[141] S. Guo, L. Wang, B. Jiang, *RSC Adv.* **2022**, *12*, 8611–8616.

[142] P. S. Singh, P. M. Badani, R. M. Kamble, *J. Photochem. Photobiol. A: Chem.* **2021**, *419*.

[143] S. Gong, X. He, Y. Chen, Z. Jiang, C. Zhong, D. Ma, J. Qin, C. Yang, *J. Mater. Chem.* **2012**, *22*, 2894–2899.

[144] N. C. Giebink, S. R. Forrest, *Phys. Rev. B* **2008**, *77*, 235215.

[145] A. K. Gupta, W. Li, A. Ruseckas, C. Lian, C. L. Carpenter-Warren, D. B. Cordes, A. M. Z. Slawin, D. Jacquemin, I. D. W. Samuel, E. Zysman-Colman, *ACS Appl. Mater. Interfaces* **2021**, *13*, 15459–15474.

[146] T. Huang, X. Song, M. Cai, D. Zhang, L. Duan, *Mater. Today Energy* **2021**, *21*.

[147] M. Einzinger, T. Zhu, P. de Silva, C. Belger, T. M. Swager, T. Van Voorhis, M. A. Baldo, *Adv. Mater.* **2017**, *29*.

[148] H. S. Kim, J. Y. Lee, S. Shin, W. Jeong, S. H. Lee, S. Kim, J. Lee, M. C. Suh, S. Yoo, *Adv. Funct. Mater.* **2021**, *31*.

[149] M. Uejima, T. Sato, M. Detani, A. Wakamiya, F. Suzuki, H. Suzuki, T. Fukushima, K. Tanaka, Y. Murata, C. Adachi, H. Kaji, *Chem. Phys. Lett.* **2014**, *602*, 80–83.

[150] H. J. Park, S. H. Han, J. Y. Lee, *Chem. Asian J.* **2017**, *12*, 2494–2500.

[151] Y. Lee, S.-J. Woo, J.-J. Kim, J.-I. Hong, *Org. Electron.* **2020**, *78*.

245

[152] P. S. Gribanov, Y. D. Golenko, M. A. Topchiy, L. I. Minaeva, A. F. Asachenko, M. S. Nechaev, *Eur. J. Org.* **2018**, *2018*, 120–125.

[153] G. Zhou, D. Song, X. Yang, Y. Zhang (Xi'an Jiaotong University), CN110818738, **2020**.

[154] S. Kato, Y. Nonaka, T. Shimasaki, K. Goto, T. Shinmyozu, *J. Org. Chem.* **2008**, *73*, 4063–4075.

[155] S. Kuila, S. Garain, G. Banappanavar, B. C. Garain, D. Kabra, S. K. Pati, S. J. George, *J. Phys. Chem. B* **2021**, *125*, 4520–4526.

[156] S. Okumura, Y. Takeda, K. Kiyokawa, S. Minakata, *Chem. Commun.* **2013**, *49*, 9266–9268.

[157] L. Hanaineh-Abdelnour, B. A. Salameh, *Heterocycles* **1999**, *51*, 2931.

[158] M. Augustin, M. Koehler, R. Harzer, G. Bernhard, H. Brigsne, *Wissenschaftliche Zeitschrift - Martin-Luther-Universitaet Halle-Wittenberg, Mathematisch-Naturwissenschaftliche Reihe* **1972**, *21*, 137–138.

[159] Q. Zhang, H. Kuwabara, W. J. Potscavage, Jr., S. Huang, Y. Hatae, T. Shibata, C. Adachi, *J. Am. Chem. Soc.* **2014**, *136*, 18070–18081.

[160] G. R. Fulmer, A. J. M. Miller, N. H. Sherden, H. E. Gottlieb, A. Nudelman, B. M. Stoltz, J. E. Bercaw, K. I. Goldberg, *Organometallics* **2010**, *29*, 2176–2179.

[161] Z. Y. Tian, X. X. Ming, H. B. Teng, Y. T. Hu, C. P. Zhang, *Chem. Eur. J.* **2018**, *24*, 13744–13748.

[162] M. Schönberger, M. Althaus, M. Fronius, W. Clauss, D. Trauner, *Nat. Chem.* **2014**, *6*, 712–719.

[163] H. Wang, L. Zhou, Y.-Z. Shi, X.-C. Fan, J.-X. Chen, K. Wang, J. Yu, X.-H. Zhang, *Chem. Eng. J.* **2022**, *433*.

[164] J. X. Chen, W. W. Tao, W. C. Chen, Y. F. Xiao, K. Wang, C. Cao, J. Yu, S. Li, F. X. Geng, C. Adachi, C. S. Lee, X. H. Zhang, *Angew. Chem. Int. Ed.* **2019**, *131*, 14802–14807.

[165] F.-M. Xie, X.-Y. Zeng, J.-X. Zhou, Z.-D. An, W. Wang, Y.-Q. Li, X.-H. Zhang, J.-X. Tang, *J. Mater. Chem. C* **2020**, *8*, 15728–15734.

[166] L. He, X. Zeng, W. Ning, A. Ying, Y. Luo, S. Gong, *Molecules* **2021**, *26*.

[167] G. Gritzner, J. Kuta, *Pure Appl. Chem.* **1984**, *56*, 461–466.

[168] N. G. Connelly, W. E. Geiger, *Chem. Rev.* **1996**, *96*, 877–910.

[169] M. Thelakkat, H.-W. Schmidt, *Adv. Mater.* **1998**, *10*, 219–223.

[170]   A. J. Bard, L. R. Faulkner, *Electrochemical Methods: Fundamentals and Applications*, John Wiley & Sons, New York, USA, **1980**.

[171]   G. M. Sheldrick, *Acta Crystallogr. A Found. Adv.* **2015**, *71*, 3–8.

[172]   G. M. Sheldrick, *Acta Crystallogr. C Struct. Chem.* **2015**, *71*, 3–8.

8. Appendix

8.1. Supplementary Data

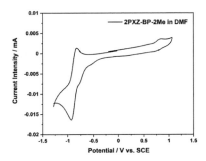

Figure 58. Cyclic voltammogram of **2PXZ-BP-2Me** measured in Ar-saturated DMF containing 0.1 M [nBu$_4$N][PF$_6$] at a scan rate of 100 mV s^{-1}.

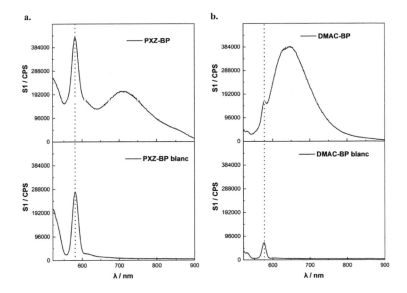

Figure 59. Fluorescence spectroscopy data for **PXZ-BP** and **DMAC-BP** measured in 0.1 mM toluene solution shown with their respective toluene blanc measurements.

248

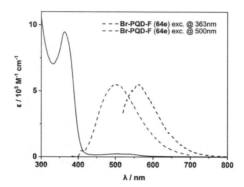

Figure 60. UV-vis spectroscopy data (in solid black line) and normalized fluorescence spectroscopy data (in dotted lines) for **Br-PQD-F (64e)** at different excitation wavelengths in 0.1 mM toluene solution.

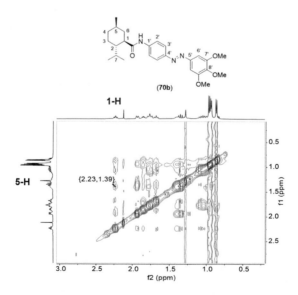

Figure 61. NOESY NMR of **Azo-Menthol-3,4,5OMe (70b)**.

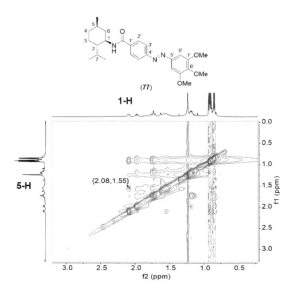

Figure 62. NOESY spectrum of **Azo-Menthyl-*N*-amide-[3,4,5]OMe (77)**.

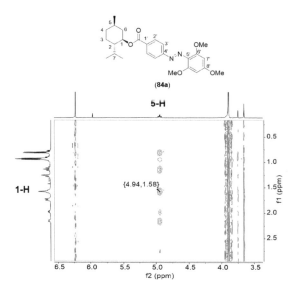

Figure 63. NOESY NMR of **Azo-Menthyl-*O*-ester-[2,4,6]OMe (84a)**.

250

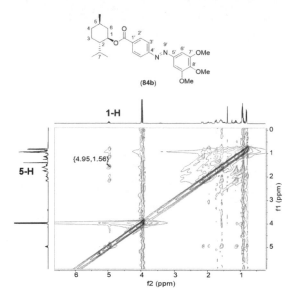

Figure 64. NOESY NMR of **Azo-Menthyl-*O*-ester-[3,4,5]OMe (84b)**.

8.2. Curriculum Vitae

Gloria Hong

gloria.hong@cdi.eu

Education

Collège des Ingénieurs (CDI)	Munich, Germany
MBA Fellow.	01/2020 – present
Science and Management Program (3-year program in parallel to Ph.D. studies).	

Karlsruhe Institute of Technology (KIT)	Karlsruhe, Germany
Ph.D. Candidate in Organic Chemistry (Prof. Dr. Bräse).	08/2019 – 08/2022

Technical University of Munich (TUM)	Garching, Germany
M.Sc. Chemistry, (1.2).	10/2016 – 04/2019
Focus areas: Organic Chemistry, Chemistry of Macromolecules, Colloids and Interfaces.	

Master Thesis Project (Prof. Dr. Süssmuth, TU Berlin), (1.0).	Berlin, Germany
"Thiyl-Radical-Induced Peptide Cyclization. Towards the Synthesis of Lipolanthines.	09/2018 – 04/2019

Technical University of Munich (TUM)	Garching, Germany
B.Sc. Chemistry, (2.0)	10/2013 – 09/2016
Bachelor Thesis Project (Prof. Dr. Gasteiger), (1.0).	
"Catalytic Effects of Transition Metal Ions on Electrolyte Reduction in Li-Ion Batteries."	

Research and Teaching Experience

New York University (NYU)	New York City, USA
Visiting Student at Trauner Group	09 – 12/2021

Karlsruhe Institute of Technology (KIT)	Karlsruhe, Germany
Graduate Researcher at Bräse Group	08/2019 – 08/2022

- Supervision of undergraduate students and two apprentices.
- Collaboration with RTG 2039 members, Wöll Group (KIT), Wenzel Group (KIT), Knebel Group (University of Jena) and Zysman-Colman Group (University of St. Andrews).

Seminar Leader	10/2020 – 02/2021

- Teaching the "Organic Chemistry II Seminar" for 90-100 undergraduate students.

Technion Israel Institute of Technology	Haifa, Israel
Research Intern at Ilan Marek Group	04 – 05/2018

- Optimization of Ti(IV)-mediated intermolecular hydroxycyclopropanation.

Technical University of Munich (TUM)	Garching, Germany
Research Intern at the WACKER Chair of Macromolecular Chemistry	11 – 12/2017

- Synthesis of *bis*-(2,2,6,6)-tetramethylpiperidin-4-yl) vinyl phosphate.

Research Intern at the Biomimetic Catalysis Group	08 – 10/2017

- Mechanistic studies of I(III)-catalyzed reactions of α-substituted cinnamic acid-derived imides.

BMW Group	Munich, Germany
Research Intern	03 – 05/2017

- Preparation and electrochemical investigation of sulfide-based solid electrolyte layers.

Technical University of Munich (TUM)	Garching, Germany
Student Research Assistant at the Chair of Technical Electrochemistry	06 – 12/2016

- Research on organic additives for Si-based electrodes in Li-ion batteries.

Teaching Fellow	10/2015 – 02/2016

- Teaching "Mathematical Methods in Chemistry I" for undergraduate students.

252

Entrepreneurial and International Experience

Technion Israel Institute of Technology   Haifa, Israel
TU Munich Exchange Program.   03 – 07/2018
Start-Up Project "Quivre" as part of the course "Business Plan for Commercializing Technology".
- Leading an international team.
- Developing a start-up idea of an electronic laboratory documentation system for chemical laboratories.
- Final pitch in front of investors.

Enactus Munich e.V.   Munich, Germany
Social Entrepreneurship.   04/2017 – 03/2018
- Team BEEautiful: Product optimization for natural cosmetics based on bee wax.
- Team reStove: Head of Social Media.

Yonsei University   Seoul, South Korea
Language Course at the Korean Language Institute.   09 – 12/2012
- Advanced writing, reading, introduction to debating.

Scholarships

KHYS Research Travel Grant   09 – 12/2021
German Academic Scholarship Foundation   04/2017 – 04/2019
e-fellows.net   09/2013 – 04/2019
Hanns-Seidel-Foundation   09/2013 – 06/2017

Extracurricular Activities

Research Training Group 2039   03/2020 – 11/2021
Ph.D. Student Representative
- Organizing soft skill seminars.
- Representing Ph.D. student interests.

TUM Department of Chemistry Faculty Council   10/2015 – 09/2016
Student Representative

TUM Student Council   10/2014 – 09/2015
Student Representative for the Department of Chemistry

8.3. List of Publications

Peer-reviewed Publications

G. Hong, C. Si, A. K. Gupta, C. Bizzarri, M. Nieger, I. D. W. Samuel, E. Zysman-Colman, S. Bräse, *J. Mater. Chem. C* **2022**, *10*, 4757-4766.

G. Hong, X. Gan, C. Leonhardt, Z. Zhang, J. Seibert, J. M. Busch, S. Bräse, *Adv. Mater.* **2021**, *33*, 2005630.

S. Solchenbach, **G. Hong**, A. T. S. Freiberg, R. Jung, H. A. Gasteiger, *J. Electrochem. Soc.* **2018**, *165*, A3304.

Conference Posters

G. Hong, C. Si, A. K. Gupta, C. Bizzarri, M. Nieger, I. D. W. Samuel, E. Zysman-Colman, S. Bräse "*Fluorinated Dibenzo[a,c]-phenazine-Based Green to Red Thermally Activated Delayed Fluorescent OLED Emitters*" from May 08th – 13th, 2022, Honolulu, Hawai'i, USA, 2022 MRS Spring Meeting & Exhibit.

G. Hong, S. Bräse "*Synthesis of Deep-Red/Near-Infrared TADF Emitters as Dyes for Cell Imaging*" from March 4th – 6th, 2020, Freudenstadt, Germany, GRK Retreat: Bioconjugation.

254

8.4. Acknowledgments

I express my deepest gratitude to Prof. Dr. Stefan Bräse for his guidance and supervision during my doctoral studies, and for giving me the opportunity to freely explore and develop new skills on a scientific and personal level, also through the CDI, my stay at NYU and various conferences.

Furthermore, I thank Prof. Dr. Eli Zysman-Colman for his valuable contribution and input on all TADF-related projects. From his group, I would like to thank Changfeng Si and David Hall for the DFT calculation result they have provided, as well as Dr. Abhishek Kumar Gupta for his work on the OLED devices for the 2D-BP-F project.

Moreover, I would like to thank Prof. Dr. Dirk Trauner for giving me the opportunity to conduct research in his group at New York University (NYU). It was a truly extraordinary time. I would like to thank all of my colleagues from the Trauner Group, but especially Alexander Sailer, Peter Rühmann, Dr. Bruno Paz, Zisis Peitsinis, Christopher Arp and Tom Ko, for the warm welcome, support and scientific discussions.

I thank Dr. Claudia Bizzarri for her support and guidance in regards to the cyclic voltammetry experiments and for giving me unrestricted excess to the necessary device and materials. She also greatly contributed to the 2D-BP-F project in which Dr. Martin Nieger was also involved. I would like to thank him for his extensive efforts on the crystal structure data, as well as his contribution and input to this project.

Dr. Alexander Knebel continues to inspire a pleasant and productive teamwork amongst collaboration partners and I would like to thank him and Bahram Hosseini Monjezi for their work on the TADF-MOF project. I thank Simon Oßwald for providing material for the TADF-linker project. Furthermore, I would like to thank Prof. Dr. Wolfgang Wenzel, Dr. Mariana Kozlowska and Anna Mauri for the collaboration on DFT calculations for TADF linkers for MOFs.

I would like to thank the GRK2039 for the financial support and the opportunity to connect and network with fellow Ph.D. students from multidisciplinary scientific backgrounds. It was my pleasure to serve as the Ph.D. student representative. Moreover, the GRK2039 has enabled the attendance of the MRS Material Research Society 2022 Spring Meeting in Honolulu, Hawai'i, through their financial support for which I am grateful.

The Karlsruhe House of Young Scientists (KHYS) played a decisive role in financing my stay at NYU for which I am very grateful and I would like to mention Mrs. Gaby Weick's efforts and support on these matters.

One of the most memorable experiences during my doctoral studies was the teamwork on the review article. I would like to thank Jasmin Seibert, Xuemin Gan, Céline Leonhardt, Zhen Zhang and Jasmin M. Busch for electing me as their team lead, and their efforts and hard work. The late-night calls and MS Teams meetings to meet deadlines, and the actual realization of what a good team can achieve in a short time will be remembered fondly!

I would like to thank Christoph Zippel for his support, and the constructive scientific discussion as well as enjoyable conversations. Also, I thank Jasmin Seibert for proof-reading this work, providing support and fun conversations between lab hours, continuous scientific discussions, hopping on the idea of a TADF literature club instantly, and many more good memories. I also want to thank Céline Leonhardt and Lisa-Lou Gracia for proof-reading part of this work. Furthermore, I want to thank Lisa-Lou Gracia, Peter Gödtel and those who have enriched my Ph.D. experience through shared memories. I would also like to mention my fellow CDI fellows Hannes Kühner and Steffen Otterbach for the joint journey until now.

I would like to acknowledge Michelle Karsten, Arina Belov and Felix Bösch for their support and work in the laboratory, and Jin San Kim for his contributions to the BP project within the framework of his Bachelor thesis project.

This work is the result of a collective effort to which the IOC Analysis Department greatly contributed. I would especially like to thank Angelika Mösle for her efforts on the mass spectrometry and infrared spectroscopy measurements, Lara Hirsch for the ESI measurements, and Despina Savvidou for the measurements by means of UV-vis spectroscopy.

Because of COVID-19, many Ph.D. students and graduates during the 2020-2021 time period have had to deal with setbacks and cancelled plans during the lockdowns and because of travel restrictions. While I am sure that each one of us will find our way nonetheless, I would like to acknowledge the unrealized achievements of those who could not attain them due to the given

256

circumstances, and the immense efforts of the PIs who navigated their research groups through this challenging time.

As this work concludes my academic career, I would like to express my gratitude to those who have taught me the methods of organic synthesis and electrochemistry, and those who have supervised my laboratory projects including Prof. Dr. Hubert A. Gasteiger (TU Munich), Prof. Dr. Ilan Marek (Technion Israel Institute of Technology), and Prof. Dr. Roderich Süssmuth (TU Berlin).

Lastly, I would like to thank my family for being who they have been and are now, and for supporting me like they have and continue to do.

For from him and
through him and
for him are all things.
To him be the glory forever.
Amen.

Romans 11:36 (NIV)